QUANTITIES, UNITS, AND SYMBOLS

A REPORT BY

THE SYMBOLS COMMITTEE OF THE ROYAL SOCIETY

REPRESENTING

THE ROYAL SOCIETY
THE CHEMICAL SOCIETY
THE INSTITUTE OF PHYSICS

Second edition

ISBN 0 85403 071 9

Approved by the Council of the Royal Society 6 March 1975

1975

PUBLISHED BY THE ROYAL SOCIETY

6 Carlton House Terrace, London SW1Y 5AG

CONTENTS

CONTENTS

PREFACE TO THE SECOND EDITION

When the 1971 edition of this booklet was published it was hoped that it would be of interest to all who were concerned with the ordering 'of all things by measure, by number, and by weight'. The complete disposal of that edition and the amount of informed comment about it showed that the beneficial effects of the standardization of units and symbols have been very widely appreciated in universities, colleges, schools, industry, government and scientific circles. The 1975 edition supersedes the 1971 edition in many respects, but the small differences of detail between them were made with deliberation by the Symbols Committee.

The Committee in preparing the urgently needed new edition resolved that it had the duty to provide an authoritative and convenient statement of the best up to date practice for the use of procedures, symbols and units by scientists in general as well as those engaged in furthering the wider range of interest of the Royal Society and other learned scientific societies. It decided that, rather than give a multiplicity of detailed references to the sources of its recommendations, it would be wiser to amplify the 1971 bibliography and to add broad statements of these sources. Indeed, it would be difficult to attempt to indicate the sources of all its individual recommendations, since selection of sources was inevitable and some recommendations resulted from logical extension of long established and internationally recognized principles. In other words, the Committee hoped to achieve a distillate of primary sources prepared with care and supported by the judgement of its members and their intimate knowledge of the work of the relevant international bodies, such as would ensure the high confidence of the scientific community in this 1975 report.

1. INTRODUCTION

The value of a *physical quantity* is equal to the product of a *numerical value* and a *unit*:

$$\text{physical quantity} = \text{numerical value} \times \text{unit}.$$

Neither any physical quantity nor the symbol used to denote it should imply a particular choice of unit.

Operations on equations involving physical quantities, units, and numerical values, should follow the ordinary rules of algebra.

Thus the physical quantity called the wavelength λ of one of the yellow sodium lines has the value

$$\lambda = 5.896 \times 10^{-7}\,\text{m}$$

where m is the symbol for the unit of length called the metre (see §3). This may equally well be written in the form

$$\lambda/\text{m} = 5.896 \times 10^{-7}$$

or in any of the other ways of expressing the equality of λ and 5.896×10^{-7} multiplied by m. By definition (see §3)

$$1\,\text{Å} = \text{Å} = 10^{-10}\,\text{m}$$

and

$$1\,\text{in} = \text{in} = 2.54 \times 10^{-2}\,\text{m}$$

where Å and in are the symbols for the units of length called respectively the ångström and the inch; it follows that

$$\lambda/\text{Å} = (\lambda/\text{m}) \times (\text{m}/\text{Å}) = 5896$$

and

$$\lambda/\text{in} = (\lambda/\text{m}) \times (\text{m}/\text{in}) = 5.896 \times 10^{-7}/(2.54 \times 10^{-2}) \approx 2.321 \times 10^{-5}.$$

Thus λ may be equated to $5.896 \times 10^{-7}\,\text{m}$, or to 5896 Å, or to $2.321 \times 10^{-5}\,\text{in}$, but may not be equated to 5.896×10^{-7} or to any other number.

It follows from the above discussion that when numerical values of a physical quantity are tabulated, the expression to be placed at the head of a column should be a pure number, such as the quotient of the symbol for the physical quantity and the symbol for the unit used.

In the following table T denotes thermodynamic temperature and K the unit of thermodynamic temperature called the kelvin. Expressions such as 'T(K)' or 'T, K' do not denote T divided by K and should be abandoned in favour of 'T/K' or '$T\ \text{K}^{-1}$'.

Example:

T/K	$10^3\,K/T$	p/MPa	ln (p/MPa)	$V_m^g/cm^3\,mol^{-1}$	pV_m^g/RT
216.55	4.6179	0.5180	−0.6578	3177.6	0.9142
273.15	3.6610	3.4853	1.2486	456.97	0.7013
304.19	3.2874	7.3815	1.9990	94.060	0.2745

Similarly, when numerical values of a physical quantity are plotted on a graph, the expression chosen to label the axis should be a pure number, such as the quotient of the symbol for the physical quantity and the symbol for the unit used.

Example:

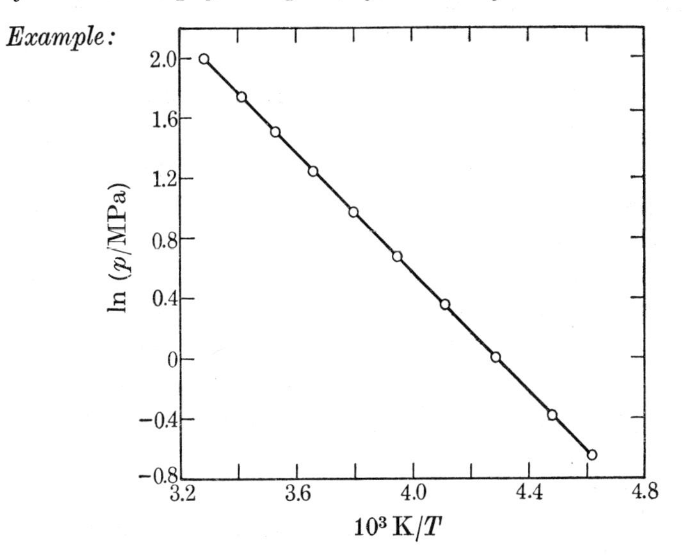

While it is desirable to maintain the above principle, its use should be flexible. For example, algebraically equivalent forms such as 'kK/T', '$(10^{-3}\,T/\mathrm{K})^{-1}$' or '1000 K/(thermodynamic temperature)' may of course be used in place of '$10^3\,\mathrm{K}/T$'.

When the graph is regarded as showing the physical quantity itself (rather than its numerical value), '$1/T$' or 'reciprocal of thermodynamic temperature' may, for example, be used as abscissa label provided that '$4.8\,\mathrm{kK}^{-1}$' is used in place of the number '4.8' on the abscissa scale.

A clear distinction should be drawn between physical quantities and units, and between the symbols for physical quantities and the symbols for units.

Symbols for physical quantities should be printed in italic (sloping) type. Symbols for units should be printed in roman (upright) type. In typescript the distinction should be made by underlining symbols for physical quantities in accord with standard printers' practice (see the Bibliography, §14.5.5 or §14.6.2).

Physical quantities and the symbols for physical quantities are dealt with in §2. The symbols for physical quantities specified there are recommendations.

Units and symbols for units are dealt with in §3. The orthography for the symbols for the units specified there is mandatory. Numbers are dealt with in §4.

2. PHYSICAL QUANTITIES AND SYMBOLS FOR PHYSICAL QUANTITIES

2.1. *Physical quantities*

Each physical quantity is given a name and a symbol which is an abbreviation for that name.

By international convention, seven physical quantities are chosen for use as dimensionally independent *base quantities*:

PHYSICAL QUANTITY	SYMBOL FOR QUANTITY
length	l
mass	m
time	t
electric current	I
thermodynamic temperature	T
amount of substance	n
luminous intensity	I_{v}

All other physical quantities are regarded as being *derived* from the base quantities.

Plane angle and solid angle are sometimes regarded as base quantities.

2.2. *Symbols for physical quantities*

The symbol for a physical quantity should be a single letter of the latin or the greek alphabet.

An exception to this rule has been made for certain dimensionless quantities used in the study of transport processes, for which the internationally agreed symbols consist of two letters, the first a capital and the second lower case. *Example:* Reynolds number: *Re*. When it is necessary to avoid ambiguity such symbols should be enclosed in parentheses.

When necessary the symbol for a physical quantity may be modified by attaching to it subscripts and/or superscripts and/or other modifying signs having specified meanings.

2.3. *Printing of symbols for physical quantities*

When letters of the latin alphabet are used as symbols for physical quantities they should be printed in italic (sloping) type. When letters of the greek alphabet are used as symbols for physical quantities they should whenever possible be printed in sloping rather than upright type.

The symbols for vector quantities should be printed in bold faced italic type. *Examples:* force: $\boldsymbol{F}$; electric field strength: $\boldsymbol{E}$. (When the directional character of such quantities is not to be emphasized, the use of ordinary italic type remains as an alternative. However, the use of bold faced italic type will often remain convenient in order to allow the use of the same letters for other quantities.)

The symbols for tensors of the second rank should be printed in bold faced sans serif type which whenever possible should be sloping rather than upright. *Examples:* ***S***, ***T***.

Abbreviations, i.e. shortened forms of names such as p.f. for partition function, should not be used in mathematical equations. When used in text they should be printed in roman (upright) type. (See also § 11.)

Recommendations on the appropriate marking of copy for the printer are given in the British Standard listed in the Bibliography, § 14.5.5. Symbols for physical quantities might also be similarly underlined in typescript or manuscript or on the blackboard.

2.4. *Choice of symbols for physical quantities*

A list of recommended symbols for physical quantities is given in §2.10. Whenever possible the symbol used for a physical quantity should be that (or one of those) recommended there.

Even with the use of both capital and lower case letters, and of bold faced as well as ordinary italic (sloping) type as specified in § 2.3, the available distinctive letter symbols are insufficient to enable each symbol to be allotted to a single quantity. Some alternatives are therefore given in the list in §2.10 where a need for them is most likely to arise or, occasionally, where alternative usages are firmly established and unobjectionable. In some instances a preference is expressed (see heading of § 2.10) and the preferred symbol should then be used whenever possible; in others no preference is expressed.

Where it is necessary to choose from alternative symbols for a quantity, or to adopt a symbol for a quantity not listed in § 2.10, consideration should be given to current practice by authorities in the field and to the desirability that symbols for quantities constituting a well defined class should as far as possible belong to the same alphabet, fount, and case.

In order to obtain additional flexibility, capital letters may be used as variants for lower case letters, and vice versa, if no ambiguity is likely to arise. For example, instead of d_i and d_e for internal and external diameter, d and D may be used. The recommended symbol for length is l and for inductance L, but l and L may also be used for two lengths or two inductances; if length and inductance appear together, however, l should be used only for length and L for inductance, and necessary distinctions between different lengths or between different inductances should be made by means of subscripts or other modifying signs.

2.5. *Modifying signs*

Letter symbols, numbers, or other signs, may be placed as subscripts or superscripts immediately after the symbol for a physical quantity in order to modify its meaning. A list of recommended symbols for some of the most commonly needed subscripts and superscripts is given in § 2.9.

For the use of other subscripts and superscripts, and of other modifying signs, no rigid rules are laid down but a satisfactory notation should fulfil the following requirements:

(i) it should be unambiguous;
(ii) it should be simple, systematic, and easy to remember;
(iii) it should not use more letters than necessary;
(iv) it should not be too expensive or difficult to print.

Modifying signs such as dots, bars, or tildes (~) may be placed above (or exceptionally below) the symbol for a physical quantity. Such signs, however, should be used sparingly and should never be letters of the alphabet or numbers.

Brackets, including parentheses ( ), braces { }, square brackets [ ], and angle brackets ⟨ ⟩, should not be used around the symbol for a quantity in order to make it represent any other quantity, unless such use is consistently adopted for a whole class of quantities as in crystallography. However, the use of square brackets around a chemical formula to denote the concentration of the substance is recommended.

2.6. *Printing of subscripts and superscripts*

Letter symbols used as subscripts or superscripts which are themselves symbols for physical quantities, and also running suffixes or exponents, should be printed in italic (sloping) type. All other letter symbols, and all numbers, used as subscripts or superscripts should be printed in roman (upright) type.

Examples: y_n for the nth value of y, but
y_2 for the 2nd value of y;
y^n for y raised to the power n, but
y^2 for y squared;
C_p for heat capacity at constant pressure, but
C_{B} for heat capacity of substance B.

When two or more subscripts, or two or more superscripts, having separate meanings are attached to the same symbol they should be separated by commas.

Example: $C_{p,\mathrm{B}}$ for heat capacity at constant pressure of substance B.

Second-order superscripts or subscripts should be avoided as far as possible. Thus e^{x^2} may be printed as $\exp x^2$. Also $\Lambda_{\mathrm{NO_3^-}}$ may be printed as $\Lambda(\mathrm{NO_3^-})$ and $\rho_{20\,°\mathrm{C}}$ as $\rho(20\,°\mathrm{C})$.

2.7. *Use of the words 'specific' and 'molar'*

The word 'specific' before the name of an extensive physical quantity should be restricted to the meaning 'divided by mass'. For example, specific volume is the volume divided by the mass. When the extensive quantity is represented by a capital letter, the corresponding specific quantity may be represented by the corresponding lower case letter.

Examples: volume: V — specific volume: $v = V/m$
heat capacity: C_p — specific heat capacity: $c_p = C_p/m$.

The numerical value of a specific physical quantity depends on the units selected for the physical quantity and for the mass.

The word 'molar' before the name of an extensive quantity should be restricted to the meaning 'divided by amount of substance'. For example, molar volume is the volume divided by the amount of substance. The subscript m attached to the symbol for the extensive quantity denotes the corresponding molar quantity.

Examples: volume: V molar volume: $V_{\mathrm{m}} = V/n$
Gibbs function: G molar Gibbs function: $G_{\mathrm{m}} = G/n$.

The subscript m may be omitted only when there is no risk of ambiguity.

The numerical value of a molar physical quantity depends on the units selected for the physical quantity and for the amount of substance (see §3.2).

2.8. *Partial molar quantities*

The symbol X_{B}, where X denotes an extensive quantity and B is the chemical symbol for a substance, denotes the partial molar quantity for the substance B defined by the relation:

$$X_{\mathrm{B}} = (\partial X/\partial n_{\mathrm{B}})_{T,p,n_{\mathrm{C}},\ldots}.$$

The partial molar quantity X_{B} for a pure substance B, which is identical with the molar quantity X_{m} for the pure substance B, may be denoted by X_{B}^{*}, where the superscript * denotes 'pure', so as to distinguish it from the partial molar quantity X_{B} for the substance B in a mixture.

2.9. *List of recommended subscripts and superscripts and other modifying signs to be used with the symbols for physical quantities*

(*a*) *Subscripts*

I, II ... 1, 2 ...	*especially with symbols for thermodynamic functions*, referring to different systems or different states of a system
A, B ...	referring to molecular species A, B...
i	referring to a typical ionic species i
u	referring to an undissociated molecule
p, V, T, S	indicating constant pressure, volume, temperature, entropy
p, m, c, a	*with symbol for an equilibrium constant*, indicating that it is expressed in terms of pressure, molality, concentration, or relative activity
g, l, s, c	referring to gas, liquid, solid, and crystalline states respectively
f, e, s, t, d	referring to fusion, evaporation, sublimation, transition, and dissolution or dilution respectively
c	referring to the critical state or indicating a critical value
C, D, F	*with symbols for optical properties*, referring to particular wavelengths
+, −	referring to a positive or negative ion, or to a positive or negative electrode
∞	indicating limiting value at infinite dilution

Some of the above subscripts may sometimes be more conveniently used as superscripts.

(b) Superscripts

$\ominus$	standard (in chemical thermodynamics)
*	indicating a pure substance
id	ideal
E	excess

2.10. *List of recommended symbols for physical quantities*

It is recognized that according to context some departures from the recommended symbols may be necessary. Where two or more symbols separated by commas are given for a quantity, these symbols are regarded as alternatives for which no preference is expressed; where they are separated by a dotted line, the first is preferred.

[*Note*: A, B denotes no preference; $A...B$ denotes A preferred]

(*a*) Space and time

angle (plane angle)	$\alpha, \beta, \gamma, \theta, \phi$, etc.	spherical coordinates	r, θ, ϕ
solid angle	Ω, ω	position vector; radius vector	$\boldsymbol{r}$
length	l	area	$A...S$
breadth	b	volume	$V...v$
height	h	time	t
thickness	d, δ	angular speed: $\mathrm{d}\theta/\mathrm{d}t$	ω
radius	r	angular acceleration: $\mathrm{d}\omega/\mathrm{d}t$	α
diameter: $2r$	d	speed: $\mathrm{d}s/\mathrm{d}t$	u, v, w
distance along path	s, L	acceleration: $\mathrm{d}u/\mathrm{d}t$	a
generalized coordinate	q	acceleration of free fall	g
rectangular coordinates	x, y, z	speed of light in a vacuum	c
cylindrical coordinates	r, ϕ, z	Mach number	Ma

(*b*) Periodic and related phenomena

period	T	circular wavenumber: $2\pi\sigma$	k
relaxation time [1]	τ	circular wavevector	$\boldsymbol{k}$
frequency: $1/T$	ν, f	damping coefficient [3]	δ
rotational frequency	n	logarithmic decrement [3]: δ/ν	Λ
circular frequency [2]: $2\pi\nu$	ω	attenuation coefficient [4]	α
wavelength	λ	phase coefficient [4]	β
wavenumber : $1/\lambda$	$\sigma ... \tilde{\nu}$	propagation coefficient [4]: $\alpha + \mathrm{i}\beta$	γ
wavevector	$\boldsymbol{\sigma}$		

[1] When F is a function of time t given by $F(t) = A + B\exp(-t/\tau)$; τ is also called time constant.

[2] Also called pulsatance.

[3] When F is a function of time t given by $F(t) = A\exp(-\delta t)\sin\{2\pi\nu(t-t_0)\}$.

[4] When F is a function of distance x given by $F(x) = A\exp(-\alpha x)\cos\{\beta(x-x_0)\}$.

(*c*) Mechanics

mass	m
density (mass density): m/V	ρ
relative density: ρ_2/ρ_1	d
specific volume: V/m	v
reduced mass: $m_1 m_2/(m_1+m_2)$	μ
momentum: mu	p
momentum (vector): $m\boldsymbol{u}$	$\boldsymbol{p}$
angular momentum	b, p_θ
angular momentum (vector): $\boldsymbol{r} \times \boldsymbol{p}$	$\boldsymbol{L}$
moment of inertia [1]	I, J
second moment of area [2]	I_a
second polar moment of area [3]	I_p
force	F
force (vector)	$\boldsymbol{F}$
weight	$G...P, W$
bending moment	M
moment of force (vector): $\boldsymbol{r} \times \boldsymbol{F}$	$\boldsymbol{M}$
torque; moment of a couple	$\boldsymbol{T}$
pressure	$p...P$
normal stress [4]	σ
shear stress [4]	τ
linear strain [4]: $\Delta l/l_0$	ϵ, e
shear strain [4]: $\Delta x/d$	γ
volume strain [4]: $\Delta V/V_0$	θ
Young modulus: σ/ϵ	E
shear modulus: τ/γ	G
bulk modulus: $-p/\theta$	K
Poisson ratio	μ, ν
compressibility: $-V^{-1}\mathrm{d}V/\mathrm{d}p$	κ
section modulus	Z, W
coefficient of friction	$\mu...f$
viscosity (dynamic viscosity)	$\eta...\mu$
fluidity: $1/\eta$	ϕ
kinematic viscosity: η/ρ	ν
diffusion coefficient	D
surface tension	γ, σ
angle of contact	θ
work	$W...A$
energy	$E...W$
potential energy	E_p, V, Φ
kinetic energy	E_k, T, K
power	P
Hamiltonian function	H
Lagrangian function	L
gravitational constant	G
Reynolds number: $\rho ul/\eta$	Re

(*d*) Thermodynamics (see also §(***h***))

thermodynamic temperature	$T...\Theta$
common temperature	t, θ
linear expansivity: $l^{-1}\mathrm{d}l/\mathrm{d}T$	α, λ
cubic expansivity: $V^{-1}\mathrm{d}V/\mathrm{d}T$	α, γ
heat; quantity of heat	$Q...q$
work; quantity of work	$W...w$
heat flow rate	$\Phi...q$
thermal conductivity	$\lambda...k$
thermal diffusivity: $\lambda/\rho c_p$	a
heat capacity	C
specific heat capacity: C/m	c
specific heat capacity at constant pressure	c_p
specific heat capacity at constant volume	c_V
ratio c_p/c_V	γ, κ
entropy	S
internal energy	$U...E$
enthalpy: $U+pV$	H
Helmholtz function: $U-TS$	A, F
Gibbs function: $U+pV-TS$	G
Massieu function: $-A/T$	J
Planck function: $-G/T$	Y

(continued)

[1] $I_z = \int (x^2+y^2)\,\mathrm{d}m$. [2] $I_{a,y} = \iint x^2\,\mathrm{d}x\mathrm{d}y$. [3] $I_p = \iint (x^2+y^2)\,\mathrm{d}x\mathrm{d}y$.

[4] More generally, stress and strain are each treated as a tensor, and a distinct notation is used.

Quantity	Symbol
specific entropy: S/m	s
specific internal energy: U/m	$u \ldots e$
specific enthalpy[1]: H/m	h
specific Helmholtz function: A/m	a, f
specific Gibbs function: G/m	g
Joule–Thomson coefficient: $(\partial T/\partial P)_H$	μ, μ_{JT}
isothermal compressibility: $-V^{-1}(\partial V/\partial p)_T$	κ, κ_T
isentropic compressibility: $-V^{-1}(\partial V/\partial p)_S$	κ_S
isobaric expansivity: $V^{-1}(\partial V/\partial T)_p$	α
thermal diffusion ratio	k_T
thermal diffusion factor	α_T
thermal diffusion coefficient	D_T

(*e*) Electricity and magnetism[2]

Quantity	Symbol
electric charge; quantity of electricity	Q
electric current: $\mathrm{d}Q/\mathrm{d}t$	I
charge density: Q/V	ρ
surface charge density: Q/A	σ
electric field strength	$\boldsymbol{E}$
electric potential	V, ϕ
electric potential difference	$U \ldots V$
electromotive force	E
electric displacement	$\boldsymbol{D}$
electric flux	Ψ
capacitance	C
permittivity ($\boldsymbol{D} = \epsilon \boldsymbol{E}$)	ϵ
electric constant; permittivity of a vacuum	ϵ_0
relative permittivity[3]: ϵ/ϵ_0	ϵ_{r}
electric susceptibility: $\epsilon_{\mathrm{r}} - 1$	χ_{e}
electric polarization: $\boldsymbol{D} - \epsilon_0 \boldsymbol{E}$	$\boldsymbol{P}$
electric dipole moment	$\boldsymbol{p} \ldots \boldsymbol{\mu}$
electric current density	$\boldsymbol{J}, \boldsymbol{j}$
magnetic field strength	$\boldsymbol{H}$
magnetic potential difference	U_{m}
magnetomotive force: $\oint H_s \mathrm{d}s$	F_{m}
magnetic flux	Φ
magnetic flux density; magnetic induction	$\boldsymbol{B}$
magnetic vector potential	$\boldsymbol{A}$
self inductance	L
mutual inductance	M, L
coupling coefficient: $L_{12}/(L_1 L_2)^{\frac{1}{2}}$	k
leakage coefficient: $1 - k^2$	σ
permeability ($\boldsymbol{B} = \mu \boldsymbol{H}$)	μ
magnetic constant; permeability of a vacuum	μ_0
relative permeability: μ/μ_0	μ_{r}
magnetic susceptibility: $\mu_{\mathrm{r}} - 1$	$\kappa \ldots \chi_{\mathrm{m}}$
magnetic moment ($\boldsymbol{T} = \boldsymbol{m} \times \boldsymbol{B}$)	$\boldsymbol{m}$
magnetization: $(\boldsymbol{B}/\mu_0) - \boldsymbol{H}$	$\boldsymbol{M}$
magnetic polarization: $\boldsymbol{B} - \mu_0 \boldsymbol{H}$	$\boldsymbol{J}$
electromagnetic energy density	w
Poynting vector: $\boldsymbol{E} \times \boldsymbol{H}$	$\boldsymbol{S}$
speed of propagation of electromagnetic waves in vacuum	c
resistance	R
resistivity ($\boldsymbol{E} = \rho \boldsymbol{J}$)	ρ
conductivity: $1/\rho$	γ, σ
reluctance: U_{m}/Φ	R, R_{m}
permeance: $1/R_{\mathrm{m}}$	Λ
number of turns	N
number of phases	m
number of pairs of poles	p
loss angle	δ
phase displacement	ϕ
impedance: $R + \mathrm{i}X$	Z
reactance: $\mathrm{Im}\, Z$	X
resistance: $\mathrm{Re}\, Z$	R
quality factor: $\lvert X \rvert / R$	Q

(*continued*)

[1] For the specific enthalpy change resulting from phase transitions the term specific latent heat is still used.

[2] Correspondences between certain quantities in this table and certain other quantities that arise when non-rationalized three-quantity systems of electric and magnetic equations are used can be found in the references given in the Bibliography, § 14.2.1, Part V or § 14.4.1.

More generally, some of the quantities appearing as scalars in this table are treated as tensors.

[3] Also called dielectric constant when it is independent of $\boldsymbol{E}$.

Quantity	Symbol
admittance: $1/Z$	Y
susceptance: Im Y	B
conductance: Re Y	G
power, active	P
power, reactive	Q
power, apparent	S

(*f*) Light and related electromagnetic radiations

The same symbol is often used for a pair of corresponding radiant and luminous quantities. Subscripts e for radiant and v for luminous may be used when necessary to distinguish these quantities.

Quantity	Symbol
radiant energy	Q, Q_e
radiant flux; radiant power	$\Phi, \Phi_e ... P$
radiant intensity: $d\Phi/d\Omega$	I, I_e
radiance: $(dI/dA)\sec\theta$	L, L_e
radiant exitance: $d\Phi/dA$	M, M_e
irradiance: $d\Phi/dA$	E, E_e
emissivity	ϵ
quantity of light	Q, Q_v
luminous flux	Φ, Φ_v
luminous intensity: $d\Phi/d\Omega$	I, I_v
luminance: $(dI/dA)\sec\theta$	L, L_v
luminous exitance: $d\Phi/dA$	M, M_v
illuminance: $d\Phi/dA$	E, E_v
light exposure: $\int E\,dt$	H
luminous efficacy: Φ_v/Φ_e	K
absorption factor; absorptance: Φ_a/Φ_0	α
reflexion factor; reflectance: Φ_r/Φ_0	ρ
transmission factor; transmittance: Φ_{tr}/Φ_0	τ
linear extinction coefficient	μ
linear absorption coefficient	a
refractive index	n
refraction: $(n^2-1)\,V/(n^2+2)$	R
angle of optical rotation	α

(*g*) Acoustics

Quantity	Symbol
speed of sound	c
speed of longitudinal waves	c_l
speed of transverse waves	c_t
group speed	c_g
sound energy flux	P
sound intensity	I, J
reflexion coefficient: P_r/P_0	ρ
acoustic absorption coefficient: $1-\rho$	$\alpha ... \alpha_a$
transmission coefficient: P_{tr}/P_0	τ
dissipation coefficient: $\alpha-\tau$	δ
loudness level	L_N

For a more complete list of symbols for acoustic quantities see the Bibliography, §14.2.1, Part VII.

(*h*) Physical chemistry

Quantity	Symbol
relative atomic mass of an element ('atomic weight') [1]	A_r
relative molecular mass of a substance ('molecular weight') [1]	M_r

(*continued*)

[1] The ratio of the average mass per atom (molecule) of the natural isotopic composition of an element (the elements) to 1/12 of the mass of an atom of the nuclide ^{12}C.

Examples: $A_r(K) = 39.102$ $A_r(Cl) = 35.453$ $M_r(KCl) = 74.555$.

The concept of relative atomic or molecular mass may be extended to other specified isotopic compositions, but the natural isotopic composition is assumed unless some other composition is specified.

amount of substance	n
molar mass: m/n	M
molar volume: V/n	V_m
molar internal energy: U/n	U_m
molar enthalpy: H/n	H_m
molar heat capacity: C/n	C_m
at constant pressure: C_p/n	$C_{p,m}$
at constant volume: C_V/n	$C_{V,m}$
molar entropy: S/n	S_m
molar Helmholtz function: A/n	A_m
molar Gibbs function: G/n	G_m
(molar) gas constant	R
compression factor: pV_m/RT	Z
mole fraction of substance B	x_B
mass fraction of substance B	w_B
volume fraction of substance B	ϕ_B
molality of solute B: (n_B divided by mass of solvent)	m_B
amount-of-substance concentration[1] of solute B: n_B/V	c_B, [B]
chemical potential of substance B: $(\partial G/\partial n_B)_{T,p,n_C...}$	μ_B
absolute activity of substance B: $\exp(\mu_B/RT)$	λ_B
partial pressure of substance B in a gas mixture: $x_B^g p$	p_B
fugacity of substance B in a gas mixture: $\lambda_B \lim_{p\to 0}(x_B^g p/\lambda_B)$	f_B, p_B^*
relative activity of substance B	a_B
activity coefficient (mole fraction basis)	f_B
activity coefficient (molality basis)	γ_B
activity coefficient (concentration basis)	y_B
osmotic coefficient	$\phi ... g$
osmotic pressure	Π
surface concentration	Γ
electromotive force	E

Faraday constant	F
charge number of ion i	z_i
ionic strength: $\frac{1}{2}\Sigma_i m_i z_i^2$	I
velocity of ion i	$\boldsymbol{v}_i$
electric mobility of ion i ($\boldsymbol{v}_i = u_i \boldsymbol{E}$)	u_i
electrolytic conductivity[2] ($\boldsymbol{J} = \kappa \boldsymbol{E}$)	κ
molar conductance of electrolyte: κ/c	Λ
transport number of ion i	t_i
molar conductance of ion i: $t_i\Lambda$	λ_i
overpotential	η
exchange current density	j_0
electrokinetic potential	ζ
intensity of light	I
transmittance: I/I_0	T
absorbance[3]: $-\lg T$	A
(linear) absorption coefficient: A/l	a
molar (linear) absorption coefficient: A/lc_B	ϵ
angle of optical rotation	α
specific optical rotatory power: $\alpha V/ml$	α_m
molar optical rotatory power: $\alpha/c_B l$	α_n
molar refraction: $(n^2-1)V_m/(n^2+2)$	R_m
stoichiometric coefficient of molecules B (negative for reactants, positive for products: the general equation for a chemical reaction is $0 = \Sigma_B \nu_B B$)	ν_B
extent of reaction ($d\xi = dn_B/\nu_B$)	ξ
affinity of a reaction[4]: $-(\partial G/\partial \xi)_{T,p} = -\Sigma_B \nu_B \mu_B$	$A ... \mathscr{A}$
equilibrium constant[5]	K
degree of dissociation	α
rate of reaction: $d\xi/dt$	$\dot{\xi}, J$
rate constant of a reaction	k
activation energy of a reaction	E

(1) Formerly called molarity.

(2) Formerly called specific conductance.

(3) Formerly called optical density.

(4) The symbol ΔG for this quantity is inappropriate since Δ implies a finite increment (see p. 32).

(5) See §2.9(*a*), p. 11.

(*i*) Molecular physics

Quantity	Symbol
Avogadro constant	L, N_A
number of molecules	N
number density of molecules: N/V	n
molecular mass	m
molecular velocity	$\boldsymbol{c}, (c_x, c_y, c_z)$; $\boldsymbol{u}, (u_x, u_y, u_z)$
molecular position	$\boldsymbol{r}, (x, y, z)$
molecular momentum	$\boldsymbol{p}, (p_x, p_y, p_z)$
average velocity	$\langle \boldsymbol{c} \rangle, \langle \boldsymbol{u} \rangle, \boldsymbol{c}_0, \boldsymbol{u}_0$
average speed	$\langle c \rangle, \langle u \rangle, \bar{c}, \bar{u}$
most probable speed	$\hat{c}, \hat{u}$
mean free path	l, λ
molecular attraction energy	ϵ
interaction energy between molecules i and j	ϕ_{ij}, V_{ij}
distribution function of speeds: $N/V = \int f \mathrm{d}c_x \mathrm{d}c_y \mathrm{d}c_z$	$f(c)$
Boltzmann function	H
generalized coordinate	q
generalized momentum	p
volume in phase space	Ω
Boltzmann constant	k
$1/kT$ in exponential functions	β
partition function	Q, Z
grand partition function	Ξ
statistical weight	g
symmetry number	σ, s
dipole moment of molecule	p, μ
quadrupole moment of molecule	Θ
polarizability of molecule	α
Planck constant	h
Planck constant divided by 2π	$\hbar$
characteristic temperature	Θ
Debye temperature: $h\nu_D/k$	Θ_D
Einstein temperature: $h\nu_E/k$	Θ_E
rotational temperature: $h^2/8\pi^2 Ik$	Θ_r
vibrational temperature: $h\nu/k$	Θ_v
Stefan–Boltzmann constant: $2\pi^5 k^4/15h^3c^2$	σ
first radiation constant[1]: $2\pi hc^2$	c_1
second radiation constant: hc/k	c_2
rotational quantum number	J, K
vibrational quantum number	v

(*j*) Atomic and nuclear physics

Quantity	Symbol
nucleon number; mass number	A
atomic number; proton number	Z
neutron number: $A-Z$	N
(rest) mass of atom	m_a
unified atomic mass constant: $m_a(^{12}C)/12$	m_u
(rest) mass of electron	m_e
(rest) mass of proton	m_p
(rest) mass of neutron	m_n
elementary charge (of proton)	e
Planck constant	h
Planck constant divided by 2π	$\hbar$
Bohr radius: $h^2/\pi\mu_0 c^2 m_e e^2$	a_0
Rydberg constant: $\mu_0^2 m_e e^4 c^3/8h^3$	R_∞
magnetic moment of particle	μ
Bohr magneton: $eh/4\pi m_e$	μ_B
Bohr magneton number: μ/μ_B [2]	
nuclear magneton: $(m_e/m_p)\mu_B$	μ_N
nuclear gyromagnetic ratio[3]: $2\pi\mu/Ih$	γ
g factor	g
Larmor (angular) frequency: $eB/2m_e$	ω_L
nuclear angular precession frequency: $geB/2m_p = \gamma B$	ω_N
cyclotron angular frequency of electron: eB/m_e	ω_c
nuclear quadrupole moment	Q
nuclear radius	R

(continued)

[1] See also page 46.

[2] No internationally agreed symbol has yet been recommended, but p ... n are in use.

[3] More logically called nuclear magnetogyric ratio (see also p. 45).

orbital angular momentum quantum number	L, l_{i}
spin angular momentum quantum number	S, s_{i}
total angular momentum quantum number	J, j_{i}
nuclear spin quantum number	I, J
hyperfine structure quantum number	F
principal quantum number	n, n_{i}
magnetic quantum number	M, m_{i}
fine structure constant: $\mu_0 e^2 c/2h$	α
electron radius: $\mu_0 e^2/4\pi m_{\mathrm{e}}$	r_{e}
Compton wavelength: $h/m_{\mathrm{e}}c$	λ_{C}
mass excess: $m_{\mathrm{a}} - A m_{\mathrm{u}}$	Δ
packing fraction: $\Delta/A m_{\mathrm{u}}$	f
mean life	τ
level width: $h/2\pi\tau$	Γ
activity: $-\mathrm{d}N/\mathrm{d}t$	A
specific activity: A/m	a
decay constant: A/N	λ
half-life: $(\ln 2)/\lambda$	$T_{\frac{1}{2}}, t_{\frac{1}{2}}$
disintegration energy	Q
spin–lattice relaxation time	T_1
spin–spin relaxation time	T_2
indirect spin–spin coupling	J

(*k*) Nuclear reactions and ionizing radiations

reaction energy	Q
cross section	σ
macroscopic cross section	Σ
impact parameter	b
scattering angle	θ, ϕ
internal conversion coefficient	α
linear attenuation coefficient	μ, μ_1
atomic attenuation coefficient	μ
mass attenuation coefficient	μ_m
linear stopping power	S, S_1
atomic stopping power	S_{a}
linear range	R, R_1
recombination coefficient	α

(*l*) Quantum mechanics

complex conjugate of Ψ	Ψ^*
probability density: $\Psi^*\Psi$	P
probability current density: $(h/4\pi i m_{\mathrm{e}})(\Psi^*\nabla\Psi - \Psi\nabla\Psi^*)$	$\boldsymbol{S}$
charge density of electrons: $-eP$	ρ
electric current density of electrons: $-e\boldsymbol{S}$	$\boldsymbol{j}$
expectation value of A	$\langle A \rangle, \bar{A}$
commutator of A and B: $AB - BA$	$[A, B], [A, B]_-$
anticommutator of A and B: $AB + BA$	$[A, B]_+$
matrix element: $\int \phi_{\mathrm{i}}^*(A\phi_{\mathrm{j}})\,\mathrm{d}\tau$	A_{ij}
Hermitian conjugate of operator A	A^{H}
momentum operator in coordinate representation	$+(h/2\pi i)\nabla$
annihilation operators	a, b, α, β
creation operators	$a^\dagger, b^\dagger, \alpha^\dagger, \beta^\dagger$

(*m*) Solid state physics

quantity	symbol
fundamental translations for lattice	$\boldsymbol{a}, \boldsymbol{b}, \boldsymbol{c}$; $\boldsymbol{a}_1, \boldsymbol{a}_2, \boldsymbol{a}_3$
Miller indices	h, k, l; h_1, h_2, h_3
plane in lattice [1]	$(h\,k\,l)$; $(h_1\,h_2\,h_3)$
direction in lattice [1]	$[u, v, w]$
fundamental translations in reciprocal lattice	$\boldsymbol{a}^*, \boldsymbol{b}^*, \boldsymbol{c}^*$; $\boldsymbol{b}_1, \boldsymbol{b}_2, \boldsymbol{b}_3$
lattice vector	$\boldsymbol{R}$
lattice plane spacing	d
Bragg angle	θ
order of reflexion	n
short range order parameter	σ
long range order parameter	s
Burgers vector	$\boldsymbol{b}$
circular wavevector; propagation vector (of phonons)	$\boldsymbol{q}$
circular wavevector; propagation vector (of particles)	$\boldsymbol{k}$
effective mass of electron	m^*, m_{eff}
Fermi energy	$E_{\mathrm{F}}, \epsilon_{\mathrm{F}}$
Fermi circular wavenumber	k_{F}
work function	Φ
differential thermoelectric power	$S \ldots \Sigma$
Peltier coefficient	Π
Thomson coefficient	μ
piezoelectric coefficient (polarization/stress)	d_{mn}
characteristic (Weiss) temperature	$\Theta, \Theta_{\mathrm{W}}$
Curie temperature	T_{C}
Néel temperature	T_{N}
Hall coefficient	R_{H}

(*n*) Molecular spectroscopy [2]

quantity	symbol
quantum number	
of component of electronic orbital angular momentum vector along symmetry axis	$\Lambda, \lambda_{\mathrm{i}}$
of component of electronic spin along symmetry axis	$\Sigma, \sigma_{\mathrm{i}}$
of total electronic angular momentum vector along symmetry axis	$\Omega, \omega_{\mathrm{i}}$
of electronic spin	S
of nuclear spin	I
of vibrational mode	v
of vibrational angular momentum (linear molecules)	l
of total angular momentum (excluding nuclear spin)	J
of component of J in direction of external field	M, M_J
of component of S in direction of external field	M_S
of total angular momentum (including nuclear spin: $\boldsymbol{F} = \boldsymbol{J}+\boldsymbol{I}$)	F
of component of F in direction of external field	M_F
of component of I in direction of external field	M_I
of component of angular momentum along axis (linear and symmetric top molecules; excluding electron- and nuclear spin; for linear molecules $K = \lvert\Lambda + l\rvert$)	K

(*continued*)

[1] Braces { } and angle brackets ⟨ ⟩ are used to enclose symmetry-related sets (forms) of planes and directions respectively. Further details regarding crystallographic notation can be found in the tables listed in the Bibliography, § 14.4.3.

[2] Further details can be found in the report listed in the Bibliography, § 14.4.4.

quantum number (cont.)

of total angular momentum (linear and symmetric top molecules; excluding electron- and nuclear spin: $J = N + S$ [1]) — N

of component of angular momentum along symmetry axis (linear and symmetric top molecules; excluding nuclear spin; for linear molecules: $P = |K + \Sigma|$ [2]) — P

degeneracy of vibrational mode — d

electronic term: E_e/hc [3] — T_e

vibrational term: E_{vib}/hc — G

coefficients in expression for vibrational term for diatomic molecule:

$G = \sigma_e(v+\frac{1}{2}) - x\sigma_e(v+\frac{1}{2})^2$ — σ_e and $x\sigma_e$

coefficients in expression for vibrational term for polyatomic molecule:

$G = \Sigma_j\sigma_j(v_j + \frac{1}{2}d_j) + \frac{1}{2}\Sigma_j\Sigma_k x_{jk}(v_j + \frac{1}{2}d_j)(v_k + \frac{1}{2}d_k)$ — σ_j and x_{jk}

rotational term: E_{rot}/hc — F

moment of inertia of diatomic molecule — I

rotational constant of diatomic molecule: $h/8\pi^2 cI$ — B

principal moments of inertia of polyatomic molecule ($I_A \leqslant I_B \leqslant I_C$) — I_A, I_B, I_C

rotational constants of polyatomic molecule: $A = h/8\pi^2 cI_A$, etc. — A, B, C

total term: $T_e + G + F$ — T

2.11. *Mathematical operations on physical quantities*

Addition and subtraction of two physical quantities are indicated by

$$a+b \quad \text{and} \quad a-b.$$

Multiplication of two (scalar [4]) physical quantities may be indicated in one of the following ways:

$$ab \quad a\,b \quad a\cdot b \quad a\times b.$$

Division of one quantity by another quantity may be indicated in one of the following ways:

$$\frac{a}{b} \quad a/b \quad ab^{-1},$$

or in any of the other ways of writing the product of a and b^{-1}.

These procedures can be extended to cases where one of the quantities or both are themselves products, quotients, sums, or differences of other quantities.

Brackets should be used in accordance with the rules of mathematics. If the solidus is used to separate the numerator from the denominator and if there is any doubt where the numerator starts or where the denominator ends, brackets should be used.

[1] System of loosely coupled electrons. [2] System of tightly coupled electrons.

[3] All energies are taken here with respect to the ground state as reference level.

[4] For vector quantities see p. 33.

Examples:

EXPRESSIONS WITH A HORIZONTAL RULE	SAME EXPRESSIONS WITH A SOLIDUS
$\dfrac{a}{bcd}$	a/bcd
$\frac{2}{9}\sin kx,\ \frac{1}{2}RT$	$(2/9)\sin kx,\ RT/2$
$\dfrac{a}{b}-c$	$a/b-c$
$\dfrac{a}{b-c}$	$a/(b-c)$
$\dfrac{a-b}{c-d}$	$(a-b)/(c-d)$
$\dfrac{a}{c}-\dfrac{b}{d}$	$a/c-b/d$

Remark. It is recommended that in expressions like:

$$\sin\{2\pi(x-x_0)/\lambda\} \qquad \exp\{(r-r_0)/\sigma\}$$
$$\exp\{-V(r)/kT\} \qquad \surd(\epsilon/c^2)$$

the argument should always be placed between brackets, except when the argument is a simple product, for example: $\sin kx$, $\sin 2\pi\nu t$.

A list of recommended symbols for mathematical operators and mathematical constants will be found in §5.

3. UNITS AND SYMBOLS FOR UNITS

3.1. *The International System of Units (SI)*

The International System of Units (SI) has been established by resolutions of the General Conference on Weights and Measures (CGPM). Details of the system itself, and of the successive decisions on which it is founded, are conveniently set out in a booklet published by the International Bureau of Weights and Measures (BIPM) (see Bibliography, § 14.3.1; English translation, § 14.3.2).

The SI units are of three kinds: *base, supplementary*, and *derived*. There are seven base units (see §§ 3.2 and 3.3), one for each of the seven physical quantities: length, mass, time, electric current, thermodynamic temperature, amount of substance, and luminous intensity, which are regarded as dimensionally independent. There are two supplementary units (see § 3.4): one for plane angle and one for solid angle. The derived unit for any other physical quantity is that obtained by the dimensionally appropriate multiplication and division of the base units (see § 3.6). Seventeen of the derived units have special names and symbols (see § 3.5).

There is one and only one SI unit for each physical quantity, although a derived unit may be denoted by more than one combination of symbols. Decimal multiples of any of the units may be constructed by means of the SI prefixes (see § 3.7).

3.2. *Definitions of the SI base units*

metre: The metre is the length equal to 1 650 763.73 wavelengths in vacuum of the radiation corresponding to the transition between the levels $2p_{10}$ and $5d_5$ of the krypton-86 atom.

kilogram: The kilogram is the unit of mass; it is equal to the mass of the international prototype of the kilogram.

second: The second is the duration of 9 192 631 770 periods of the radiation corresponding to the transition between the two hyperfine levels of the ground state of the caesium-133 atom.

ampere: The ampere is that constant current which, if maintained in two straight parallel conductors of infinite length, of negligible circular cross-section, and placed 1 metre apart in vacuum, would produce between these conductors a force equal to 2×10^{-7} newton per metre of length.

kelvin: The kelvin, unit of thermodynamic temperature, is the fraction 1/273.16 of the thermodynamic temperature of the triple point of water.

mole: The mole is the amount of substance of a system which contains as many elementary entities as there are atoms in 0.012 kilogram of carbon 12. When the mole is used, the elementary entities must be specified and may be atoms, molecules, ions, electrons, other particles, or specified groups of such particles.[1]

candela: The candela is the luminous intensity, in the perpendicular direction, of a surface of 1/600 000 square metre of a black body at the temperature of freezing platinum under a pressure of 101 325 newtons per square metre.

3.3. *Names and symbols for the SI base units*

PHYSICAL QUANTITY	NAME OF SI UNIT	SYMBOL FOR SI UNIT
length	metre	m
mass	kilogram	kg
time	second	s
electric current	ampere	A
thermodynamic temperature	kelvin	K
amount of substance	mole	mol
luminous intensity	candela	cd

3.4. *Names and symbols for the SI supplementary units*

PHYSICAL QUANTITY	NAME OF SI UNIT	SYMBOL FOR SI UNIT
plane angle	radian	rad
solid angle	steradian	sr

3.5. *Special names and symbols for SI derived units*

PHYSICAL QUANTITY	NAME OF SI UNIT	SYMBOL FOR SI UNIT	DEFINITION OF SI UNIT[2]	EQUIVALENT FORM(S) OF SI UNIT[2]
energy	joule	J	m^2 kg s^{-2}	N m
force	newton	N	m kg s^{-2}	J m^{-1}
pressure	pascal	Pa	m^{-1} kg s^{-2}	N m^{-2}, J m^{-3}
power	watt	W	m^2 kg s^{-3}	J s^{-1}
electric charge	coulomb	C	s A	A s
electric potential difference	volt	V	m^2 kg s^{-3} A^{-1}	J A^{-1} s^{-1}, J C^{-1}
electric resistance	ohm	Ω	m^2 kg s^{-3} A^{-2}	V A^{-1}
electric conductance	siemens	S	m^{-2} kg^{-1} s^3 A^2	Ω^{-1}, A V^{-1}
electric capacitance	farad	F	m^{-2} kg^{-1} s^4 A^2	A s V^{-1}, C V^{-1}
magnetic flux	weber	Wb	m^2 kg s^{-2} A^{-1}	V s
inductance	henry	H	m^2 kg s^{-2} A^{-2}	V A^{-1} s

(*continued*)

[1] Examples of the use of the mole:
1 mole of HgCl has a mass equal to 0.236 04 kilogram.
1 mole of Hg_2Cl_2 has a mass equal to 0.472 08 kilogram.
1 mole of e^- has a mass equal to 5.4860×10^{-7} kilogram.
1 mole of a mixture containing $\frac{2}{3}$ mole of H_2 and $\frac{1}{3}$ mole of O_2 has a mass equal to 0.012 010 2 kilogram.

[2] The sequence of unit symbols within a compound unit is optional.

PHYSICAL QUANTITY	NAME OF SI UNIT	SYMBOL FOR SI UNIT	DEFINITION OF SI UNIT	EQUIVALENT FORM(S) OF SI UNIT
magnetic flux density	tesla	T	$kg\ s^{-2}\ A^{-1}$	$V\ s\ m^{-2}$, $Wb\ m^{-2}$
luminous flux	lumen [1]	lm	cd sr	
illuminance	lux [1]	lx	m^{-2} cd sr	
frequency	hertz	Hz	s^{-1}	
activity (of a radioactive source; nuclear transformations per unit time)	becquerel [2]	Bq	s^{-1}	
absorbed dose (of ionizing radiation)	gray [2]	Gy	$J\ kg^{-1}$	

3.6. ***Examples of SI derived units and unit symbols for other quantities***

(This list is merely illustrative)

PHYSICAL QUANTITY	SI UNIT	A SYMBOL FOR SI UNIT
area	square metre	m^2
volume	cubic metre	m^3
wavenumber	1 per metre	m^{-1}
density	kilogram per cubic metre	$kg\ m^{-3}$
speed; velocity	metre per second	$m\ s^{-1}$, $m{\cdot}s^{-1}$
angular velocity	radian per second	$rad\ s^{-1}$
acceleration	metre per second squared	$m\ s^{-2}$, $m{\cdot}s^{-2}$
kinematic viscosity	square metre per second	$m^2\ s^{-1}$
amount-of-substance concentration	mole per cubic metre	$mol\ m^{-3}$
specific volume	cubic metre per kilogram	$m^3\ kg^{-1}$
molar volume	cubic metre per mole	$m^3\ mol^{-1}$
dynamic viscosity	pascal second	Pa s
moment of force	newton metre	N m
surface tension	newton per metre	$N\ m^{-1}$
heat flux density	watt per square metre	$W\ m^{-2}$
heat capacity	joule per kelvin	$J\ K^{-1}$
thermal conductivity	watt per metre kelvin	$W\ m^{-1}\ K^{-1}$
energy density	joule per cubic metre	$J\ m^{-3}$
molar heat capacity	joule per kelvin mole	$J\ K^{-1}\ mol^{-1}$
electric field strength	volt per metre	$V\ m^{-1}$

(*continued*)

(1) In the definition given here for these units, the steradian (sr) is treated as a base unit.

(2) The special names and symbols for these two units were sanctioned by the 15th CGPM (1975) instead of the curie and the rad (see page 28).

PHYSICAL QUANTITY	SI UNIT	A SYMBOL FOR SI UNIT
magnetic field strength	ampere per metre	$A\ m^{-1}$
magnetic moment	joule per tesla	$J\ T^{-1}$
electric charge density	coulomb per cubic metre	$C\ m^{-3}$
permittivity	farad per metre	$F\ m^{-1}$
current density	ampere per square metre	$A\ m^{-2}$
permeability	henry per metre	$H\ m^{-1}$
luminance	candela per square metre	$cd\ m^{-2}$

3.7. *SI prefixes*

The following prefixes may be used to construct decimal multiples of units.

MULTIPLE	PREFIX	SYMBOL	MULTIPLE	PREFIX	SYMBOL
10^{-1}	deci	d	10	deca	da
10^{-2}	centi	c	10^{2}	hecto	h
10^{-3}	milli	m	10^{3}	kilo	k
10^{-6}	micro	μ	10^{6}	mega	M
10^{-9}	nano	n	10^{9}	giga	G
10^{-12}	pico	p	10^{12}	tera	T
10^{-15}	femto	f	10^{15}	peta	P [1]
10^{-18}	atto	a	10^{18}	exa	E [1]

Decimal multiples of the kilogram (kg) should be formed by attaching an SI prefix not to kg but to g, in spite of the kilogram and not the gram being the SI base unit.

Examples: mg not μkg for 10^{-6} kg
Mg not kkg for 10^{3} kg.

A symbol for an SI prefix may be attached to the symbol for an SI base unit (§ 3.3), or for an SI supplementary unit (§ 3.4), or for an SI derived unit having a special name and symbol (§3.5).

Examples: cm ns μA mK μmol μrad MHz daN kPa GV MΩ.

SI prefixes may be attached to one or more of the unit symbols in an expression for a compound unit.

Examples: $V\ ms^{-1}$ $\mu mol\ dm^{-3}$.

An SI prefix is also sometimes attached to the symbol for a non-SI unit.

Compound prefixes should not be used.

Example: nm but not mμm for 10^{-9} m.

A combination of prefix and symbol for a unit is regarded as a single symbol which may be raised to a power without the use of brackets.

Example: cm^2 always means $(0.01\ m)^2$ and never $0.01\ m^2$.

[1] The names and symbols for these two prefixes were sanctioned by the 15th CGPM (1975).

3.8. *Units recognized for continued use together with SI*

These units are not part of SI but it is recognized that they will continue to be used in appropriate contexts (e.g. day and year are convenient units of time for events with diurnal or annual periodicity).

SI prefixes may be attached to some of these units, e.g. millilitre, ml; mega-electronvolt, MeV.

It is recommended that compound units formed from these units with SI units be used only in limited cases (e.g. for purely descriptive purposes when values of the quantity will not be substituted into algebraic equations).

Except for the unit M or M (see footnote (4) to table (*a*)), all of the units in the following two tables are recognized by CIPM for continued use with SI.

(*a*) *Units exactly defined in terms of SI units*

PHYSICAL QUANTITY	NAME OF UNIT	SYMBOL FOR UNIT	DEFINITION OF UNIT
time	minute	min	60 s
time	hour	h	60 min = 3600 s
time	day	d	24 h = 86400 s
angle	degree	°	$(\pi/180)$ rad
angle	minute	′	$(\pi/10800)$ rad
angle	second	″	$(\pi/648\,000)$ rad
volume	litre [(1)]	l	10^{-3} m^3 = dm^3
mass	tonne	t	10^3 kg = Mg
Celsius temperature (t_C) [(2)]	degree Celsius	°C [(3)]	K
amount-of-substance concentration	— [(4)]	M, M	10^3 mol m^{-3} = mol dm^{-3}

(1) By decision of the twelfth General Conference of Weights and Measures in October 1964 the old definition of the litre (leading to the approximate value 1.000028 dm^3) was rescinded and the word litre reinstated as a special name for the cubic decimetre. Neither the word litre nor its symbol l should be used to express results of high precision.

(2) The Celsius temperature is the excess of the thermodynamic temperature over 273.15 K.

(3) The ° sign and the letter following form one symbol and there should be no space between them. Example: 25 °C not 25° C.

(4) The name 'molar' for this unit should be avoided because of the danger of confusion with the definition of 'molar' quantities (§ 2.7). The unit should be avoided in precise work because of the change in the definition of the litre, and because of confusion with molality. Until a name is given to the SI unit for concentration (mol m^{-3}), however, the use of M (or M) is likely to persist for stating concentrations in aqueous solutions in situations where the precision is no higher than, say, $1/10^3$.

(b) Units defined in terms of certain physical constants

PHYSICAL QUANTITY	NAME OF UNIT	SYMBOL FOR UNIT	CONVERSION
length	astronomical unit	AU [1]	1 AU $\approx 149\,600 \times 10^{6}$ m
length	parsec	pc	1 pc $\approx 30\,857 \times 10^{12}$ m
mass	unified atomic mass unit	u	1 u $\approx 1.660\,565\,5 \times 10^{-27}$ kg
energy	electronvolt	eV	1 eV $\approx 1.602\,189\,2 \times 10^{-19}$ J
time	year [2]	a	—

3.9. *Other units*

It is recognized that some of these units will continue in use for a limited time, but the Symbols Committee recommend that their use in scientific publications be abandoned as quickly as practicable. (For the electromagnetic units of the CGS system see § 3.11.)

(a) Decimal multiples of SI units having special names

PHYSICAL QUANTITY	NAME OF UNIT	SYMBOL FOR UNIT	DEFINITION OF UNIT
length	ångström	Å	10^{-10} m = 10^{-1} nm
length	micron	μm [3]	10^{-6} m
area	are	are [4]	10^{2} m^{2}
area	barn	b	10^{-28} m^{2}
force	dyne	dyn	10^{-5} N
pressure	bar	bar	10^{5} Pa
energy	erg	erg	10^{-7} J
kinematic viscosity	stokes	St	10^{-4} m^{2} s^{-1}
dynamic viscosity	poise	P	10^{-1} Pa s
acceleration of free fall	gal	Gal	1 cm s^{-2} = 10^{-2} m s^{-2}
magnetic flux density	gamma	γ	10^{-9} T
luminance	stilb	sb	1 cd cm^{-2} = 10^{4} cd m^{-2}
illuminance	phot	ph	10^{4} lx

(b) Units exactly defined in terms of SI units

These units are not part of SI. It is recognized that their use may be continued for some time but it is recommended that except in special circumstances they should be progressively abandoned in scientific publications. These units should not be used to form compound units with SI units. The following list is by no means exhaustive. Each of the definitions given in the fourth column is *exact*.

[1] This unit has no international symbol. AU is the abbreviation of the English name; the abbreviations of the French and German names are UA and AE respectively.

[2] The year is not uniquely defined.

[3] The symbols μ and mμ should give place to μm (micrometre) and nm (nanometre) respectively.

[4] The full name is shown here in place of any unit symbol.

PHYSICAL QUANTITY	NAME OF UNIT	SYMBOL FOR UNIT	DEFINITION OF UNIT
length	inch	in	2.54×10^{-2} m
mass	pound (avoirdupois)	lb	0.453 592 37 kg
force	kilogram-force	kgf	9.806 65 N
energy	thermochemical calorie	cal_{th}	4.184 J
energy	I.T. calorie	cal_{IT}	4.1868 J
pressure	atmosphere	atm	101 325 Pa
pressure	torr	Torr	(101 325/760) Pa [≈ 133.322 368 Pa]
pressure	conventional millimetre of mercury	mmHg	$13.5951 \times 980.665 \times 10^{-2}$ Pa [≈ 133.322 387 Pa]
thermodynamic temperature (T)	degree Rankine	°R [1]	$\frac{5}{9}$K
Fahrenheit temperature (t_F) [2]	degree Fahrenheit	°F [1]	$\frac{5}{9}$K
activity [3]	curie	Ci	3.7×10^{10} Bq = 3.7×10^{10} s^{-1}
absorbed dose [3]	rad	rad [4]	10^{-2} Gy = 10^{-2} J kg^{-1}
exposure [3]	röntgen	R	2.58×10^{-4} C kg^{-1}

3.10. *'International' electric units*

These units are obsolete, having been replaced by the 'absolute' (SI) units in 1948. The conversion factors which should be used with electric measurements quoted in 'international' units depend on where and when the instruments used to make the measurements were calibrated. The following two sets of conversion factors refer respectively to the 'mean international' units estimated by the ninth General Conference of Weights and Measures in 1948, and to the 'U.S. international' units estimated by the U.S. National Bureau of Standards as applying to instruments calibrated by them before 1948.

1 'mean international ohm' = 1.000 49 Ω

1 'mean international volt' = 1.000 34 V

1 'U.S. international ohm' = 1.000 495 Ω

1 'U.S. international volt' = 1.000 330 V

(1) The ° sign and the letter following it form one symbol and there should be no space between them. *Example:* 25 °F not 25° F.

(2) The Fahrenheit temperature is the excess of the thermodynamic temperature over 459.67 °R.

(3) Information on other radiation quantities and units may be found in ICRU report no. 19, part I (see § 14.4.5). Now that special names and symbols have been given to the corresponding SI units (see page 24), the curie and the rad will become obsolete.

(4) Whenever confusion with the symbol for the radian (angular measure) appears possible the symbol rd may be used.

3.11. *Electric and magnetic units belonging to unit systems other than SI*

Definitions of units used in the 'electrostatic CGS' and 'electromagnetic CGS' unit systems can be found in either of two documents listed in the Bibliography, §14.2.1, Part V or §14.4.1.

It appears that for many years to come a knowledge of the 'electromagnetic CGS' unit system will be a necessity for workers in magnetism, but for practical purposes it is usually sufficient to note that 1 maxwell (Mx) corresponds to 10^{-8} Wb, that 1 gauss (G) corresponds to 10^{-4} T, and that 1 oersted (Oe) corresponds to $10^3(4\pi)^{-1}\,\mathrm{A\,m^{-1}} \approx 79.5775\,\mathrm{A\,m^{-1}}$.

3.12. *Printing of symbols for units*

The symbol for a unit should be printed in roman (upright) type, should remain unaltered in the plural, and should not be followed by a full stop except when it occurs at the end of a sentence.

Example: 5 cm but not 5 cms. and not 5 cm. and not 5 cms

The symbol for a unit derived from a proper name should begin with a capital roman (upright) letter.

Examples: J for joule and Hz for hertz.

Any other symbol for a unit should be printed in lower case roman (upright) type.

Symbols for prefixes for units should be printed in roman (upright) type with no space between the prefix and the unit.

3.13. *Multiplication and division of units*

A product of two units may be represented in either of the ways:

$$\mathrm{N\,m} \quad \text{or} \quad \mathrm{N\cdot m}$$

The representation Nm is not recommended.

A quotient of two units may be represented in any of the ways:

$$\mathrm{m/s} \quad \text{or} \quad \frac{\mathrm{m}}{\mathrm{s}} \quad \text{or} \quad \mathrm{m\,s^{-1}} \quad \text{or} \quad \mathrm{m\cdot s^{-1}}$$

but not $\mathrm{ms^{-1}}$.

Where there is danger of confusion it is wise to use $\mathrm{m\cdot N}$, $\mathrm{m\cdot s^{-1}}$, $\mathrm{m\cdot s^{-2}}$ instead of $\mathrm{m\,N}$, $\mathrm{m\,s^{-1}}$, $\mathrm{m\,s^{-2}}$ lest these latter be mistaken for mN, $\mathrm{ms^{-1}}$, $\mathrm{ms^{-2}}$.

These rules may be extended to more complex groupings, but more than one solidus (/) should not be used in the same expression unless parentheses are used to eliminate ambiguity.

Examples: $\mathrm{J\ K^{-1}\,mol^{-1}}$ or $\mathrm{J/(K\ mol)}$ but not $\mathrm{J/K/mol}$

$\mathrm{cm^2\,V^{-1}\,s^{-1}}$ or $\mathrm{(cm/s)/(V/cm)}$ but not $\mathrm{cm/s/V/cm}$.

4. NUMBERS

4.1. *Printing of numbers*

Numbers should normally be printed in upright type. The decimal sign between digits in a number should be a point (.) or a comma (,). To facilitate the reading of long numbers the digits may be grouped in threes about the decimal sign but no point or comma should ever be used except for the decimal sign.

Example: 2 573.421 736 but not 2,573.421,736.

When the decimal sign is placed before the first digit of a number a zero should always be placed before the decimal sign.

Example: 0.2573×10^4 but not $.2573 \times 10^4$.

It is often convenient to print numbers with just one digit before the decimal sign.

Example: 2.573×10^3.

4.2. *Multiplication and division of numbers*

The multiplication sign between numbers should be a cross (×).

Example: 2.3×3.4.

Division of one number by another may be indicated in any of the ways:

$$\frac{136}{273} \quad \text{or} \quad 136/273 \quad \text{or} \quad 136 \times (273)^{-1}.$$

These rules may be extended to more complex groupings, but more than one solidus (/) should never be used in the same expression unless parentheses are used to eliminate ambiguity.

Example: $(136/273)/2.303$ or $136/(273 \times 2.303)$ but never $136/273/2.303$.

5. RECOMMENDED MATHEMATICAL SYMBOLS

[*Note*: A, B denotes no preference; A...B denotes A preferred]

equal to	$=$	smaller than	$<$
not equal to	$\neq$	larger than	$>$
identically equal to	$\equiv$	smaller than or equal to	$\leqslant$
corresponds to	$\triangleq$	larger than or equal to	$\geqslant$
approximately equal to	$\approx$	much smaller than	$\ll$
approaches	$\rightarrow$	much larger than	$\gg$
asymptotically equal to	$\simeq$, $\sim$	plus	$+$
proportional to	$\propto$	minus	$-$
infinity	∞	plus or minus	$\pm$
		minus or plus	$\mp$

a multiplied by b (1)	ab, $a \cdot b$, $a \times b$
a divided by b (1)	a/b, $\frac{a}{b}$, ab^{-1}
magnitude of a	$\|a\|$
a raised to power n	a^n
square root of a	$a^{\frac{1}{2}}$, $\sqrt{a}$
nth root of a	$a^{1/n}$, $a^{\frac{1}{n}}$, $\sqrt[n]{a}$
mean value of a	$\langle a \rangle$, $\bar{a}$
sign of a ($a/\|a\|$ for $a \neq 0$)	$\mathrm{sgn}\, a$
greatest integer less than or equal to a	$\mathrm{ent}\, a$
integer part of a	$[a]$
factorial p (2)	$p!$
binomial coefficient (3)	$\binom{n}{p}$

When letters of the alphabet are used to form mathematical operators (*Examples:* d, Δ, ln, exp) or as mathematical constants (*Examples:* e, π) they should be printed in roman (upright) type so as to distinguish them from the symbols for physical quantities which should be printed in italic (sloping) type.

sum	Σ
product	Π
function of x	$f(x)$, $\mathrm{f}(x)$
limit to which $f(x)$ tends as x approaches a	$\lim_{x \to a} f(x)$, $\lim_{x \to a} f(x)$

(*continued*)

(1) See also §2.11.

(2) $p! = 1 \times 2 \times 3 \times \ldots \times (p-1) \times p$ where p is a positive integer.

(3) $\binom{n}{p} = n!/[(n-p)!\,p!]$ where n and p are positive integers and $n \geqslant p$ and where $0! = 1$.

finite increment of x	Δx
variation of x	δx
differential coefficient of $f(x)$ with respect to x	$\frac{\mathrm{d}f}{\mathrm{d}x}$, $\mathrm{d}f/\mathrm{d}x$, $f'(x)$
differential coefficient of order n of $f(x)$	$\frac{\mathrm{d}^n f}{\mathrm{d}x^n}$, $\mathrm{d}^n f/\mathrm{d}x^n$, $f^{(n)}(x)$
partial differential coefficient of $f(x, y, \ldots)$ with respect to x when $y, \ldots$ are held constant	$\frac{\partial f(x,y,\ldots)}{\partial x}$, $\left(\frac{\partial f}{\partial x}\right)_y$, $(\partial f/\partial x)_y$, f_x
operator $\frac{\partial}{\partial x}$ or with single variable $\frac{\mathrm{d}}{\mathrm{d}x}$	D_x, D
the total differential of f	$\mathrm{d}f$
indefinite integral of $f(x)$ with respect to x	$\int f(x)\,\mathrm{d}x$
definite integral of $f(x)$ from $x = a$ to $x = b$	$\int_a^b f(x)\,\mathrm{d}x$
integral of $f(x)$ with respect to x round a closed contour	$\oint f(x)\,\mathrm{d}x$
convolution product, integral of $f(x)\,g(y-x)$ from $x = 0$ to $x = y$	$f * g$
exponential of x	$\exp x$, e^x
base of natural logarithms	e
logarithm to the base a of x	$\log_a x$
natural logarithm of x	$\ln x$, $\log_\mathrm{e} x$
common logarithm of x	$\lg x$, $\log_{10} x$
binary logarithm of x	$\mathrm{lb}\, x$, $\log_2 x$
ratio of circumference to diameter of a circle	π
sine of x	$\sin x$
cosine of x	$\cos x$
tangent of x	$\tan x$
cotangent of x	$\cot x$
secant of x	$\sec x$
cosecant of x	$\mathrm{cosec}\, x$
inverse sine[1] of x	$\arcsin x \ldots \sin^{-1} x$
inverse cosine of x	$\arccos x \ldots \cos^{-1} x$
inverse tangent of x	$\arctan x \ldots \tan^{-1} x$
inverse cotangent of x	$\mathrm{arccot}\, x \ldots \cot^{-1} x$
inverse secant of x	$\mathrm{arcsec}\, x \ldots \sec^{-1} x$
inverse cosecant of x	$\mathrm{arccosec}\, x \ldots \mathrm{cosec}^{-1} x$
hyperbolic sine of x	$\sinh x$
hyperbolic cosine of x	$\cosh x$
hyperbolic tangent of x	$\tanh x$
hyperbolic cotangent of x	$\coth x$

(*continued*)

[1] The notation $\sin^{-1} x$, etc., is discouraged because of possible confusion with $(\sin x)^{-1}$, etc.

hyperbolic secant of x	$\operatorname{sech} x$
hyperbolic cosecant of x	$\operatorname{cosech} x$
inverse hyperbolic sine [1] of x	$\operatorname{arsinh} x \ldots \sinh^{-1} x$
inverse hyperbolic cosine of x	$\operatorname{arcosh} x \ldots \cosh^{-1} x$
inverse hyperbolic tangent of x	$\operatorname{artanh} x \ldots \tanh^{-1} x$
inverse hyperbolic cotangent of x	$\operatorname{arcoth} x \ldots \coth^{-1} x$
inverse hyperbolic secant of x	$\operatorname{arsech} x \ldots \operatorname{sech}^{-1} x$
inverse hyperbolic cosecant of x	$\operatorname{arcosech} x \ldots \operatorname{cosech}^{-1} x$
complex operator ($\mathrm{i}^2 + 1 = 0$)	i, j
real part of z	$\operatorname{Re} z$
imaginary part of z	$\operatorname{Im} z$
modulus of z	$\|z\|$
argument of z	$\arg z$
complex conjugate of z	z^*
transpose of matrix A	$\tilde{A}, A^{\mathrm{T}}$
complex conjugate matrix of matrix A	A^*
Hermitian conjugate matrix [2] of matrix A	A^{H}
pseudo (or generalized) inverse of matrix A	A^+
determinant of matrix A	$\det A$
trace of matrix A	$\operatorname{tr} A$
vector	$\boldsymbol{A} \ldots \vec{A}$
magnitude of vector $\boldsymbol{A}$	$\|\boldsymbol{A}\|, A$
scalar (or inner) product of vectors $\boldsymbol{A}$ and $\boldsymbol{B}$	$\boldsymbol{A} \cdot \boldsymbol{B}$
vector product of vectors $\boldsymbol{A}$ and $\boldsymbol{B}$	$\boldsymbol{A} \times \boldsymbol{B}, \boldsymbol{A} \wedge \boldsymbol{B}$
dyadic (or outer) product of vectors $\boldsymbol{A}$ and $\boldsymbol{B}$	$\boldsymbol{A}\boldsymbol{B}$
differential vector operator	$\nabla, \dfrac{\partial}{\partial \boldsymbol{r}}$
gradient of ϕ	$\operatorname{grad} \phi, \nabla \phi$
divergence of $\boldsymbol{A}$	$\nabla \cdot \boldsymbol{A}, \operatorname{div} \boldsymbol{A}$
curl of $\boldsymbol{A}$	$\nabla \times \boldsymbol{A}, \nabla \wedge \boldsymbol{A}, \operatorname{curl} \boldsymbol{A}, \operatorname{rot} \boldsymbol{A}$
Laplacian of ϕ: $\nabla \cdot \nabla \phi$	$\nabla^2 \phi$
Dalembertian of ϕ: $\nabla^2 \phi - c^{-2}\, \partial^2 \phi / \partial t^2$	$\Box \phi$
double inner product of tensors $\boldsymbol{S}$ and $\boldsymbol{T}$	$\boldsymbol{S} : \boldsymbol{T}$
single inner product of tensors $\boldsymbol{S}$ and $\boldsymbol{T}$	$\boldsymbol{S} \cdot \boldsymbol{T}$
outer product of tensors $\boldsymbol{S}$ and $\boldsymbol{T}$	$\boldsymbol{S}\boldsymbol{T}$
inner product of tensor $\boldsymbol{S}$ and vector $\boldsymbol{A}$	$\boldsymbol{S} \cdot \boldsymbol{A}$

[1] The notation $\sinh^{-1} x$, etc., is discouraged because of possible confusion with $(\sinh x)^{-1}$, etc.

[2] The use of $A^\dagger$ to denote the Hermitian conjugate is discouraged because of possible confusion with A^+.

6. CHEMICAL ELEMENTS, NUCLIDES, AND PARTICLES

6.1. *Definitions*

A nuclide is a species of atoms identical as regards atomic number (proton number) and mass number (nucleon number). Two or more nuclides having the same atomic number but different mass numbers are called isotopes or isotopic nuclides. Two or more nuclides having the same mass number are called isobars or isobaric nuclides.

6.2. *Symbols for elements and nuclides*

Symbols for chemical elements should be written in roman type. The symbol is not followed by a full stop.

Examples: Ca, C, H, He.

The attached numerals specifying a nuclide are as follows:

$$\text{mass number}\ {}^{14}\mathrm{N}_{2\ \text{atoms/molecule}}$$

The atomic number may be placed in the left subscript position.

The right superscript position should be used, when required, to indicate ionic charge, state of excitation, or oxidation number.

Examples:

ionic charge: Cl^-, SO_4^{2-}, Ca^{2+}, PO_4^{3-}
electronic excited states: He^*, NO^*
nuclear excited states: ${}^{110}Ag^*$, ${}^{110}Ag^m$
oxidation number: $K_6M^{IV}Mo_9O_{32}$.

6.3. *Symbols for particles and quanta*

neutron	n	pion	π
proton	p	muon	μ
deuteron	d	electron	e
triton	t	neutrino	ν
helion	h	photon	γ
α-particle	α		

It is recommended that the following notation should be used:

Hyperons: Upright capital greek letters to indicate specific particles, e.g. Λ, Σ.

Nucleons: Upright lower case n and p to indicate neutron and proton respectively.

Mesons: Upright lower case greek letters to indicate specific particles, e.g. π, μ, τ.

Leptons: L-particles; e.g. e, ν.

It is recommended that the charge of particles be indicated by adding the superscript +, −, or 0.

Examples: π^+, π^-, π^0; p^+, p^-; e^+, e^-.

If with the symbols p and e no sign is shown then the symbols should refer to the positive proton and the negative electron respectively.

The symbol ~ above the symbol of a particle should indicate the corresponding antiparticle (e.g. $\tilde{\nu}$ for anti-neutrino).

6.4. *Notation for nuclear reactions*

The meaning of the symbolic expression indicating a nuclear reaction should be the following:

$$\text{initial nuclide}\left(\begin{array}{c}\text{incoming particle(s)}\\ \text{or quanta}\end{array}, \begin{array}{c}\text{outgoing particle(s)}\\ \text{or quanta}\end{array}\right)\text{final nuclide.}$$

Examples: $^{14}N(\alpha, p)^{17}O$ $^{59}Co(n, \gamma)^{60}Co$

$^{23}Na(\gamma, 3n)^{20}Na$ $^{31}P(\gamma, pn)^{29}Si$.

7. QUANTUM STATES

7.1. *General rules*

A letter symbol indicating the quantum state of *a system* should be printed in capital upright type. A letter symbol indicating the quantum state of *a single electron* should be printed in lower case upright type.

7.2. *Atomic spectroscopy*

The letter symbols indicating quantum states are:

$L, l = 0$: S, s	$L, l = 4$: G, g	$L, l = 8$: L, l
$= 1$: P, p	$= 5$: H, h	$= 9$: M, m
$= 2$: D, d	$= 6$: I, i	$= 10$: N, n
$= 3$: F, f	$= 7$: K, k	$= 11$: O, o.

A right hand subscript indicates the total angular momentum quantum number J or j. A left hand superscript indicates the spin multiplicity $2S + 1$.

Examples: $^{2}\mathrm{P}_{\frac{3}{2}}$ - state ($J = \frac{3}{2}$, multiplicity 2)

$\mathrm{p}_{\frac{3}{2}}$ - electron ($j = \frac{3}{2}$).

An atomic electron configuration is indicated symbolically by:

$$(nl)^{\kappa}\,(n'l')^{\kappa'}\ldots.$$

The quantum symbols s, p, d, f, ... are used instead of $l = 0, 1, 2, 3, \ldots$.

Example: the atomic configuration: $(1\mathrm{s})^2\,(2\mathrm{s})^2\,(2\mathrm{p})^3$.

7.3. *Molecular spectroscopy*

The letter symbols indicating molecular electronic quantum states are for *linear molecules*:

$$\Lambda, \lambda = 0\colon \Sigma, \sigma$$
$$= 1\colon \Pi, \pi$$
$$= 2\colon \Delta, \delta$$

and for *non-linear molecules*:

A, a; B, b; E, e; etc.

A left hand superscript indicates the spin multiplicity. For molecules having a symmetry centre the parity symbol g or u, indicating respectively symmetric or antisymmetric behaviour on inversion, is attached as a right hand subscript. A + or − sign attached as a right hand superscript indicates the symmetry as regards reflexion in any plane through the symmetry axis of the molecules.

Examples: Σ_g^+, Π_u, $^2\Sigma$, $^3\Pi$, etc.

The letter symbols indicating the vibrational angular momentum states in the case of *linear molecules* are:

$$l = 0: \Sigma$$
$$= 1: \Pi$$
$$= 2: \Delta.$$

7.4. *Nuclear spectroscopy*

The spin and parity assignment of a nuclear state is

$$J^\pi$$

where the parity symbol π is + for even and − for odd parity.

Examples: 3^+, 2^-, etc.

A shell model configuration is indicated symbolically by:

$$(n\,l\,j)^\kappa\,(n'l'j')^{\kappa'}$$

where the first bracket refers to the proton shell and the second to the neutron shell. Negative values of κ or κ' indicate holes in a completed shell. Instead of $l = 0, 1, 2, 3, \ldots$ the quantum state symbols s, p, d, f, ... are used.

Example: The nuclear configuration $(1\,\mathrm{d}\,\frac{3}{2})^3\,(1\,\mathrm{f}\,\frac{7}{2})^2$.

7.5. *Spectroscopic transitions*

The upper level and the lower level are indicated by ′ and ″ respectively.

Examples: $h\nu = E' - E''$ $\quad \sigma = T' - T''$.

A spectroscopic transition should be indicated by writing the upper state first and the lower state second, connected by a dash in between.

Examples:	$^2\mathrm{P}_{\frac{1}{2}}$—$^2\mathrm{S}_{\frac{1}{2}}$	for an electronic transition
	(J', K')—(J'', K'')	for a rotational transition
	v'—v''	for a vibrational transition.

Absorption transition and emission transition may be indicated respectively by arrows ← and →.

Examples: $(J', K') \leftarrow (J'', K'')$ absorption from (J'', K'') to (J', K')

$^2P_{\frac{1}{2}} \rightarrow {}^2S_{\frac{1}{2}}$ emission from $^2P_{\frac{1}{2}}$ to $^2S_{\frac{1}{2}}$.

The difference Δ between two quantum numbers should be that of the upper state minus that of the lower state.

Example: $\Delta J = J' - J''$.

The indications of the branches of the rotation band should be as follows:

$$\begin{aligned} \Delta J = J' - J'' &= -2: \text{O-branch} \\ &= -1: \text{P-branch} \\ &= 0: \text{Q-branch} \\ &= +1: \text{R-branch} \\ &= +2: \text{S-branch.} \end{aligned}$$

8. NUCLEAR PHYSICS

8.1. *Notation for covariant character of coupling*

S	scalar coupling	A	axial vector coupling
V	vector coupling	P	pseudoscalar coupling
T	tensor coupling		

8.2. *Character of transitions*

Multipolarity of transition:

electric or magnetic	monopole	E0 or M0
	dipole	E1 or M1
	quadrupole	E2 or M2
	octopole	E3 or M3
	2^n-pole	En or Mn

parity change in transition:

transition *with* parity change: yes

transition *without* parity change: no

8.3. *Sign of polarization vector* (*Basel convention*)

In nuclear interactions the positive polarization of particles with spin $\frac{1}{2}$ is taken in the direction of the vector product

$$\boldsymbol{k}_{\mathrm{i}} \times \boldsymbol{k}_{\mathrm{o}},$$

where $\boldsymbol{k}_{\mathrm{i}}$ and $\boldsymbol{k}_{\mathrm{o}}$ are the circular wavevectors of the incoming and outgoing particles respectively.

9. THERMODYNAMIC RESULTS

Thermodynamic results for chemical or physical processes should be expressed by quoting the equation for the process (with such specification as may be necessary of the initial and final physical states, including the composition, of the system) followed by the value of the change in the appropriate thermodynamic function.

Examples:

$H_2(g) + \frac{1}{2}O_2(g) = H_2O(l)$; $\Delta H^\ominus_m(298.15\,K) = -285.83\,kJ\,mol^{-1}$.

$2H_2(g) + O_2(g) = 2H_2O(l)$; $\Delta H^\ominus_m(298.15\,K) = -571.66\,kJ\,mol^{-1}$.

$H_2(g) + \frac{1}{2}O_2(g) = H_2O(l)$; $\Delta G^\ominus_m(298.15\,K) = -237.18\,kJ\,mol^{-1}$.

$H_2O(l) = H_2O(g)$; $\Delta H^\ominus_m(298.15\,K) = +44.01\,kJ\,mol^{-1}$.

$AgCl(s) + \frac{1}{2}H_2(g) = Ag(s) + HCl\,\{m(HCl) = 0.2857\,mol\,kg^{-1}, m(C_6H_{12}O_6\!:\textsc{d}\text{-glucose})$
$= 2.379\,mol\,kg^{-1}, aq\}$; $A(298.15\,K,\ 101.325\,kPa) = +28.11\,kJ\,mol^{-1}$.

The following symbols should be used to specify physical states. They should be printed in roman type and should be placed in parentheses after the formula of the substance as in the examples given above.

g gaseous
l liquid
s solid
c crystalline
aq dissolved at effectively infinite dilution in water.

10. GALVANIC CELLS

10.1. *The electromotive force of a cell*

The cell should be represented by a diagram, e.g.

$$Zn|Zn^{2+}|Cu^{2+}|Cu.$$

The electromotive force is equal in sign and magnitude to the electrical potential of the metallic conducting lead on the right when that of the similar lead on the left is taken as zero, the circuit being open.

When the reaction of the cell is written as

$$\tfrac{1}{2}Zn + \tfrac{1}{2}Cu^{2+} \rightarrow \tfrac{1}{2}Zn^{2+} + \tfrac{1}{2}Cu,$$

this implies a diagram so drawn that this reaction takes place when positive electricity flows through the cell from left to right. If this is the direction of the current when the cell is short-circuited, as in the present example, the electromotive force is positive.

If, however, the reaction is written as

$$\tfrac{1}{2}Cu + \tfrac{1}{2}Zn^{2+} \rightarrow \tfrac{1}{2}Cu^{2+} + \tfrac{1}{2}Zn$$

this implies the diagram $Cu|Cu^{2+}|Zn^{2+}|Zn$

and the electromotive force of the cell so specified is negative.

10.2. *The electromotive force of a half cell and the so-called 'electrode potential'*

The term 'electromotive force of a half cell' as applied to half cells written as follows:

$$Zn^{2+}|Zn$$
$$Cl^{-}|Cl_2, Pt$$
$$Cl^{-}|AgCl, Ag$$
$$Fe^{2+}, Fe^{3+}|Pt$$

means the electromotive forces of the cells:

$Pt, H_2|H^+|Zn^{2+}|Zn$	implying the reaction	$\tfrac{1}{2}H_2 + \tfrac{1}{2}Zn^{2+} \rightarrow H^+ + \tfrac{1}{2}Zn$
$Pt, H_2|H^+|Cl^-|Cl_2, Pt,$		$\tfrac{1}{2}H_2 + \tfrac{1}{2}Cl_2 \rightarrow H^+ + Cl^-$
$Pt, H_2|H^+|Cl^-|AgCl, Ag$		$\tfrac{1}{2}H_2 + AgCl \rightarrow H^+ + Cl^- + Ag$
$Pt, H_2|H^+|Fe^{2+}, Fe^{3+}|Pt$		$\tfrac{1}{2}H_2 + Fe^{3+} \rightarrow H^+ + Fe^{2+}$

where the electrode on the left is a *standard hydrogen electrodc.*

These electromotive forces may also be called *relative electrode potentials* or, in brief, *electrode potentials*.

On the other hand, the term 'electromotive force of a half cell' as applied to half cells written as follows:

$$Zn|Zn^{2+}$$
$$Pt, Cl_2|Cl^-$$
$$Ag, AgCl|Cl^-$$
$$Pt|Fe^{2+}, Fe^{3+}$$

means the electromotive forces of the cells:

$Zn|Zn^{2+}|H^+|H_2, Pt$	implying the reaction	$\frac{1}{2}Zn + H^+ \rightarrow \frac{1}{2}Zn^{2+} + \frac{1}{2}H_2$
$Pt, Cl_2|Cl^-|H^+|H_2, Pt$		$Cl^- + H^+ \rightarrow \frac{1}{2}Cl_2 + \frac{1}{2}H_2$
$Ag, AgCl|Cl^-|H^+|H_2, Pt$		$Ag + Cl^- + H^+ \rightarrow AgCl + \frac{1}{2}H_2$
$Pt|Fe^{2+}, Fe^{3+}|H^+|H_2, Pt$		$Fe^{2+} + H^+ \rightarrow Fe^{3+} + \frac{1}{2}H_2$

where the electrode on the right is a *standard hydrogen electrode*.

These electromotive forces should not be called electrode potentials.

11. SOME COMMON ABBREVIATIONS

This list is not intended to be exhaustive. The words in this list will often be given in full in the text, but where abbreviations are used the following forms are recommended. Such abbreviations should be printed in roman type (except for *ca.*).

absolute	abs.
adenosine triphosphate	ATP
adrenocorticotrophic hormone	ACTH
alternating current	a.c.
anhydrous	anhyd.
approximate(-ly)	approx., *ca.*
aqueous	aq.
atomic mass	at. mass
average	av.
boiling point	b.p.
calculated	calc.
central nervous system	c.n.s.
centre of gravity	c.g.
constant	const.
count per minute	count/min
corrected	corr.
critical	crit.
deoxyribonucleic acid	DNA
diameter, inside	i.d.
diameter, outside	o.d.
dilute	dil.
direct current	d.c.
distilled	dist.
electrocardiogram	e.c.g.
electromagnetic unit	e.m.u.
electromotive force	e.m.f.
electron spin resonance	e.s.r.
electrostatic unit	e.s.u.
ethyl	Et
experiment	expt
freezing point	f.p.
gas-liquid chromatography	g.l.c.
Greenwich Mean Time	G.M.T.
haemoglobin	Hb
infrared	i.r.
liquid	liq.
magnetomotive force	m.m.f.
maximum	max.
melting point	m.p.
methyl	Me
minimum	min.
molecular mass	molec. mass
nuclear magnetic resonance	n.m.r.
observed	obs.
parts per million	$/10^6$
per cent	% (or in full)
potential difference	p.d.
radio frequency	r.f.
recrystallized	recryst.
red blood corpuscle	r.b.c.
relative humidity	r.h.
respiratory quotient	r.q.
ribonucleic acid	RNA
root mean square	r.m.s.
section, paragraph	§
soluble	sol.
solution	soln
standard deviation	s.d.
standard error	s.e.
standard temperature and pressure	s.t.p.
thin layer chromatography	t.l.c.
Temps Universel	T.U.
ultraviolet	u.v.
Universal Time	U.T.
vacuum	vac.
vapour density	v.d.
vapour pressure	v.p.

12. RECOMMENDED VALUES OF PHYSICAL CONSTANTS

The following values are based on the set of constants recommended by the CODATA Committee (see Bibliography, §14.4.6). The standard-deviation uncertainty is given below each value. Details concerning the development of this self-consistent set of values and their uncertainties are given by Cohen & Taylor (1973); see Bibliography, §14.6.4.

QUANTITY	SYMBOL	VALUE AND STANDARD-DEVIATION UNCERTAINTY
speed of light in a vacuum	c	$2.997\,924\,580 \times 10^{8}$ m s^{-1} 12
magnetic constant, permeability of a vacuum	μ_0	$4\pi \times 10^{-7}$ H m^{-1} (exact)
electric constant, permittivity of a vacuum	$\epsilon_0 = \mu_0^{-1}c^{-2}$	$8.854\,187\,82 \times 10^{-12}$ F m^{-1} 7
fine structure constant	$\alpha = \mu_0 e^2 c/2h$	$7.297\,350\,6 \times 10^{-3}$ 60
	α^{-1}	137.036 04 11
elementary charge (of proton)	e	$1.602\,189\,2 \times 10^{-19}$ C 46
Planck constant	h	$6.626\,176 \times 10^{-34}$ J s 36
	$\hbar = h/2\pi$	$1.054\,588\,7 \times 10^{-34}$ J s 57
magnetic flux quantum	$\Phi_0 = h/2e$	$2.067\,850\,6 \times 10^{-15}$ V s 54
Avogadro constant	L, N_A	$6.022\,045 \times 10^{23}$ mol^{-1} 31
unified atomic mass constant	m_u	$1.660\,565\,5 \times 10^{-27}$ kg 86
rest mass: of electron	m_e	$9.109\,534 \times 10^{-31}$ kg 47
	m_e/m_u	$5.485\,802\,6 \times 10^{-4}$ 21
of proton	m_p	$1.672\,648\,5 \times 10^{-27}$ kg 86
	m_p/m_u	1.007 276 471 11

(continued)

QUANTITY	SYMBOL	VALUE AND STANDARD DEVIATION UNCERTAINTY
rest mass (*cont.*):		
of proton	m_p/m_e	1836.151 52 70
of neutron	m_n	$1.674\,954\,3 \times 10^{-27}$ kg 86
	m_n/m_u	1.008 665 012 37
Faraday constant	F	$9.648\,456 \times 10^{4}$ C mol^{-1} 27
Rydberg constant	$R_\infty = \mu_0^2 m_e e^4 c^3/8h^3$	$1.097\,373\,177 \times 10^{7}$ m^{-1} 83
Hartree energy	$E_H = 2R_\infty hc$	$4.359\,814 \times 10^{-18}$ J 24
Bohr radius	$a_0 = h^2/\pi\mu_0 c^2 m_e e^2$	$5.291\,770\,6 \times 10^{-11}$ m 44
electron radius	$r_e = \mu_0 e^2/4\pi m_e$	$2.817\,938\,0 \times 10^{-15}$ m 70
Bohr magneton	$\mu_B = eh/4\pi m_e$	$9.274\,078 \times 10^{-24}$ J T^{-1} 36
magnetic moment:		
of electron	μ_e	$9.284\,832 \times 10^{-24}$ J T^{-1} 36
	μ_e/μ_B	1.001 159 656 7 35
of proton	μ_p	$1.410\,617\,1 \times 10^{-26}$ J T^{-1} 55
	μ_p/μ_B	$1.521\,032\,209 \times 10^{-3}$ 16
gyromagnetic ratio of protons in H_2O	γ'_p	$2.675\,130\,1 \times 10^{8}$ s^{-1} T^{-1} 75
	$\gamma'_p/2\pi$	$4.257\,602 \times 10^{7}$ s^{-1} T^{-1} 12
γ'_p corrected for diamagnetism of H_2O	γ_p	$2.675\,198\,7 \times 10^{8}$ s^{-1} T^{-1} 75
	$\gamma_p/2\pi$	$4.257\,711 \times 10^{7}$ s^{-1} T^{-1} 12
nuclear magneton	$\mu_N = (m_e/m_p)\,\mu_B$	$5.050\,824 \times 10^{-27}$ J T^{-1} 20
magnetic moment of protons in H_2O (μ'_p)	μ'_p/μ_B	$1.520\,993\,22 \times 10^{-3}$ 10

(*continued*)

QUANTITY	SYMBOL	VALUE AND STANDARD DEVIATION UNCERTAINTY
proton magnetic moment (uncorrected) divided by nuclear magneton	μ'_p/μ_N	2.792 774 0 1 1
μ'_p/μ_N corrected for diamagnetism of H_2O	μ_p/μ_N	2.792 845 6 1 1
Compton wavelength:		
of electron	$\lambda_C = h/m_e c$	$2.426\,308\,9 \times 10^{-12}$ m 4 0
of proton	$\lambda_{C,p} = h/m_p c$	$1.321\,409\,9 \times 10^{-15}$ m 2 2
of neutron	$\lambda_{C,n} = h/m_n c$	$1.319\,590\,9 \times 10^{-15}$ m 2 2
gas constant	R	8.314 41 J K^{-1} mol^{-1} 26
Boltzmann constant	$k = R/L$	$1.380\,662 \times 10^{-23}$ J K^{-1} 44
Stefan–Boltzmann constant	$\sigma = 2\pi^5 k^4/15h^3c^2$	$5.670\,32 \times 10^{-8}$ W m^{-2} K^{-4} 71
first radiation constant (1)	$c_1 = 2\pi hc^2$	$3.741\,832 \times 10^{-16}$ J m^2 s^{-1} 20
	$8\pi hc$	$4.992\,563 \times 10^{-24}$ J m 27
second radiation constant	$c_2 = hc/k$	$1.438\,786 \times 10^{-2}$ m K 45
gravitational constant	G	$6.672\,0 \times 10^{-11}$ N m^2 kg^{-2} 4 1

Accurate values of commonly occurring mathematical constants

QUANTITY	SYMBOL	VALUE
ratio of circumference to diameter of a circle	π	3.141 592 653 59
base of natural logarithms	e	2.718 281 828 46
natural logarithm of 10	ln 10	2.302 585 092 99

(1) The spectral radiant exitance (formerly called spectral radiant emittance and sometimes emissive power), M_λ, is given by

$$M_\lambda = 2\pi hc^2\lambda^{-5}/\{\exp(hc/kT\lambda) - 1\};$$

the spectral radiant energy density, w_λ, is given by

$$w_\lambda = 8\pi hc\lambda^{-5}/\{\exp(hc/kT\lambda) - 1\};$$

unfortunately there is no accepted name or symbol for the constant $8\pi hc$.

13. SOURCES

13.1. General

The highest international authority with respect to the names, definitions, and symbols for physical quantities and with respect to the algebra of relations among quantities or among quantities, numbers, and units (including conventions about dimensions) is the International Organization for Standardardization (ISO) through its Technical Committee 12 (ISO/TC 12). International Standards are prepared by ISO/TC 12 after consultations with the national standards institutions (through which any scientist may exercise a 'national' voice) and with the International (subject) Unions (through which any scientist may exercise a 'subject' voice). In electrical and magnetic matters the International Electrotechnical Commission (IEC) enjoys similar status, but ISO and IEC work closely together and do not publish conflicting recommendations. The International Unions publish their own recommendations, but all have agreed not to make recommendations which conflict with International Standards.

The names, definitions, and symbols for units are the responsibility of the General Conference of Weights and Measures (CGPM), at least for all countries such as the UK which adhere to the Metre Convention.

The Committee on Data for Science and Technology (CODATA) of the International Council of Scientific Unions is responsible for recommending the current 'best' values of the fundamental physical constants.

13.2. Relating to individual sections

The following paragraphs give the sources relevant to individual sections of this booklet. Details of the publications to which reference is made are given in §14 (Bibliography).

§*1. Introduction.* In most respects, this follows ISO 31/0, sections B1–B3 (§14.2.1).

§*2. Physical quantities and symbols for physical quantities.* The general subsections 2.1–2.6 follow ISO 31/0 (§14.2.1).

§*2.7. Use of the words 'specific' and 'molar'*, and

§*2.8. Partial molar quantities.* These sections follow IUPAC (1970), section 1.4 (§14.4.2). They are consistent with the usage in ISO/DIS 31/III and ISO 31/VIII (§14.2.1).

§*2.9. List of recommended subscripts....* The list of subscripts follows the usage throughout ISO 31 (§14.2.1); most of the superscripts appear in IUPAC (1970), section 2.11 (§14.4.2).

§*2.10. List of recommended symbols for physical quantities.* Tables (*a*)–(*h*) and (*j*) are based primarily on the corresponding parts of the International Standard ISO 31 (e.g. table (*a*) is based on ISO/R 31/Part I; table (*f*) is based on ISO 31/VI, etc; table (*j*) is based on ISO 31/IX. Details of the published editions of these standards are given in the Bibliography (§ 14.2.1); some of the recommendations in these tables have been amended in accordance with existing drafts for revised editions of these standards.

Certain quantities in these fields are not mentioned in ISO 31 but appear in the tables of symbols recommended by the International Union of Pure and Applied Chemistry (§ 14.4.2) and the International Union of Pure and Applied Physics (§ 14.4.1). Most of these are included in tables (*a*)–(*h*).

Tables (*i*) and (*k*)–(*n*), dealing with specialized fields of physics, are based primarily on the tables with corresponding headings in the IUPAP booklet (§ 14.4.1) (table (*k*) is based on section 7.9, 'Atomic and nuclear physics', in that booklet). Most of the material in tables (*i*), (*l*) and (*n*) is not yet covered by ISO recommendations.

A few of the quantities in these tables do not appear in any of the published international documents; in these cases the recommendations conform to current usage.

In a very few instances, the recommendations in the published international documents have appeared to the Symbols Committee to be out of date, and a recommendation in conformity with current usage has been substituted.

§*2.11. Mathematical operations on physical quantities.* This section follows para. C.1.2 of ISO 31/0 (§ 14.2.1), except that the latter permits the form *a.b* for the product of two quantities in addition to those mentioned in § 2.11.

§*3. Units and symbols for units.* Almost all the recommendations in this section are derived from decisions of the CGPM or of the CIPM recorded in their publications (§ 14.1). A convenient summary of these decisions, with references to the original sources, is the booklet prepared by the BIPM (English translation: § 14.3.2).

§*3.13. Multiplication and division of units* is based on ISO 31/0, para. C.2.2 (§ 14.2.1), and ISO 1000 para. 4.5.2 (§ 14.2.2).

§*4. Numbers.* This section is based on ISO 31/0 section C.3 and ISO/R 31/Part XI (§ 14.2.1).

§*5. Recommended mathematical symbols.* This section follows ISO/R 31/Part XI (§ 14.2.1), with a few additions in conformity with current practice.

§*6.2. Symbols for elements and nuclides* follows ISO 31/0 section C.4 (§ 14.2.1) (for oxidation number: IUPAC (1970), section 7.2 (§ 14.4.2)).

§*6.3. Symbols for particles and quanta* follows IUPAP (1965), section 4 (§ 14.4.1) with a few additions.

§*6.4. Notation for nuclear reactions* follows IUPAP (1965), section 6.3 (§ 14.4.1).

§*7. Quantum states* follows IUPAP (1965) section 5 (§ 14.4.1).

§*8. Nuclear physics* follows IUPAP (1965) section 6 (§14.4.1).

§*9. Thermodynamic results* follows current practice; these matters are not yet the subject of any international agreement.

§*10. Galvanic cells* follows IUPAC (1970) section 9 (§14.4.2).

§*11. Some common abbreviations* follows current usage.

§*12. Recommended values of physical constants* is selected from CODATA (1973) (§14.4.6).

14. BIBLIOGRAPHY

This bibliography is inevitably selective rather than comprehensive.

Many of the items listed are subject to repeated revision; they should be consulted in their latest edition.

14.1. The publications of the bodies of the Metre Convention

The proceedings of the General Conference, the International Committee, the Consultative Committees, and the International Bureau are published under the auspices of the Bureau in the following series:

Comptes rendus des séances de la Conférence Générale des Poids et Mesures; Procès-Verbaux des séances du Comité International des Poids et Mesures; Sessions des Comités Consultatifs;

Recueil de Travaux du Bureau International des Poids et Mesures (this compilation brings together articles published in scientific and technical journals and books, as well as certain work issued in the form of reports).

The collection of the *Travaux et Mémoires du Bureau International des Poids et Mesures* (22 volumes published between 1881 and 1966) ceased in 1966 by a decision of the International Committee.

From time to time the International Bureau publishes a report on the development of the Metric System throughout the world, entitled *Les récents progrès du Système Métrique.*

Since 1965 the international journal *Metrologia*, edited under the auspices of the International Committee of Weights and Measures, has published articles on the principal work on scientific metrology carried out throughout the world and on the improvement in measuring methods and standards, units, etc., as well as reports concerning the activities, decisions, and recommendations of the various bodies created under the Metre Convention.

A convenient summary of the most important decisions recorded in these series, together with detailed references, is given in a booklet prepared by the Bureau International des Poids et Mesures (see § 14.3 below).

14.2. The publications of the work of Technical Committee 12 of the International Organization for Standardization (ISO/TC 12)

14.2.1. *International Standard* ISO 31

An international standard on quantities, units, symbols, conversion factors and conversion tables issues in various parts. (These were formerly published under the title *ISO Recommendations R* 31.) SI units are given precedence throughout

ISO 31, but other units are listed and defined in terms of SI units. A reference is given below to the latest version of each part, and also to the latest version of any *Draft International Standard* (DIS, formerly *Draft ISO Recommendation*, DIR) intended to supersede the last published version.

Part 0: ISO 31/0 (1974). General introduction to ISO/31. General principles concerning quantities, units and symbols.

Part I: ISO/R 31/Part I (1965); ISO/DIS 31/I (1975). Quantities and units of space and time.

Part II: ISO/R 31/Part II (1958); ISO/DIS 31/II (1975). Quantities and units of periodic and related phenomena.

Part III: ISO/R 31/Part III (1960); ISO/DIS 31/III (1975). Quantities and units of mechanics.

Part IV: ISO/R 31/Part IV (1960); ISO/DIS 31/IV (1975). Quantities and units of heat.

Part V: ISO/R 31/Part V (1965); ISO/DIS 31/V (1975). Quantities and units of electricity and magnetism.

Part VI: ISO 31/VI (1973). Quantities and units of light and related electromagnetic radiations.

Part VII: ISO/R 31/Part VII (1965); ISO/DIS 31/VII (1975). Quantities and units of acoustics.

Part VIII: ISO 31/VIII (1973). Quantities and units of physical chemistry and molecular physics.

Part IX: ISO 31/IX (1973). Quantities and units of atomic and nuclear physics.

Part X: ISO 31/X (1973). Quantities and units of nuclear reactions and ionizing radiations.

Part XI: ISO/R 31/Part XI (1961). Mathematical signs and symbols for use in physical sciences and technology.

Part XII: ISO 31/XII (1975). Dimensionless parameters.

Part XIII: ISO/DIS 31/XIII (1974). Quantities and units of solid state physics.

14.2.2. *International Standard* ISO 1000 (1973)

SI units and recommendations for the use of their multiples and of certain other units. (See also § 14.5.2.)

14.2.3. *Availability*

The ISO standards and recommendations listed above are leaflets of 6–21 pages. They may be consulted at the library of the British Standards Institution (BSI), 2 Park Street, London W1A 2BS, and they are held by the public libraries of Leeds, Liverpool and Sheffield. They may be purchased through BSI but their price (fixed by ISO) is high (average about £5 each in 1974). They may also be obtained through the inter-library loan network.

14.3. Booklet prepared by the Bureau International des Poids et Mesures

14.3.1

SI Le Système International d'Unités (2nd ed. 1973; 40 pages). Paris: OFFILIB, 48 rue Gay-Lussac, F75, Paris 5e.

14.3.2

SI The International System of Units (1973, translation of the preceding entry, prepared jointly by the National Physical Laboratory, U.K., and the National Bureau of Standards, U.S.A., and approved by BIPM; 47 pages). London: H.M.S.O.; Washington: U.S. Gov. Printing Office.

14.4. Publications of other international organizations

14.4.1

International Union of Pure and Applied Physics, S.U.N. Commission 1965, *Symbols, units and nomenclature in physics* (32 pages). Document U.I.P. 11 (S.U.N. 65–3). (New edition in preparation.)

14.4.2

International Union of Pure and Applied Chemistry, Division of Physical Chemistry, Commission on Symbols, Terminology, and Units 1970. *Manual of symbols and terminology for physicochemical quantities and units* (44 pages). London: Butterworths. Also published in *Pure and Applied Chemistry* 1970, **21**, 1–44.

14.4.3

International Union of Crystallography 1969. *International tables for X-ray crystallography* vol. 1 (558 pages). Birmingham: Kynoch Press.

14.4.4

Joint Commission for Spectroscopy of I.U.P.A.P. and I.A.U. 1955. Report on notation for the spectra of polyatomic molecules. *Journal of Chemical Physics*, **23**, 1997–2011.

14.4.5

International Commission on Radiation Units and Measurements 1971. *Radiation quantities and units*. Report no. 19, part I.

14.4.6

CODATA 1973. Recommended consistent values of the fundamental physical constants, 1973. CODATA Bulletin no. 11.

14.5 Publications of the British Standards Institution (BSI)

14.5.1

BS 3763 1970 *The International System of Units (SI)* (12 pages). This British Standard gives definitions and symbols for the SI units and prefixes, and for the units outside SI that are recognized by CIPM, with a brief introduction and historical note. It follows closely reference 14.3.2.

14.5.2

BSI PD 5686 1972 *The use of SI units* (28 pages). After a short general and historical introduction, this document reproduces ISO 1000 (see § 14.2.2) with only minor editorial changes.

14.5.3

BS 350 *Conversion factors and tables*:
Part 1 1974 Basis of tables, conversion factors (100 pages).
Part 2 1962 Detailed conversion tables (293 pages).
Supplement No. 1 1967 to BS 350, Part 2 (PD 6203) Additional tables for SI conversions (87 pages).

14.5.4

BS 1991 *Letter symbols, signs and abbreviations*:
Part 1 1967 General (45 pages).
Part 2 1961 Chemical engineering, nuclear science and applied chemistry (45 pages).
Part 3 1961 Fluid mechanics (36 pages).
Part 4 1961 Structures, materials and soil mechanics (49 pages).
Part 5 1961 Applied thermodynamics (30 pages).
Part 6 1975 Electrical science and engineering (21 pages).

14.5.5

BS 1219 1958 *Recommendations for proof correction and copy preparation* (27 pages).

14.5.6. *Availability*

BSI publications may be purchased from the British Standards Institution, Sales Branch, Newton House, 101 Pentonville Road, London N1 9ND, or across the counter for cash at the Institution's headquarters, 2 Park Street, London W1A 2BS, where they may also be consulted in its library. They may also be purchased at BSI's sales departments in Birmingham, Bristol, Glasgow and Liverpool. BSI publications may also be obtained through the inter-library loan network.

14.6. Other documents

14.6.1

McGlashan, M. L. 1971 *Physicochemical quantities and units (the grammar and spelling of physical chemistry)* (2nd ed.) (128 pages). London: The Royal Institute of Chemistry.

14.6.2

Chaundy, T. W., Barrett, P. R. & Batey, C. 1954 *The printing of mathematics* (116 pages). Oxford University Press.

14.6.3

The Royal Society 1974 *General notes on the preparation of scientific papers* (32 pages). London: The Royal Society.

14.6.4.

Cohen, E. R. & Taylor, B. N. 1973 *Journal of Physical and Chemical Reference Data* **2**, 663.

Printed in Great Britain at the University Printing House, Cambridge
(Euan Phillips, University Printer)

NOTES